The PWS
BookWare Companion Series™

Labs for Signals and Systems

Using MATLAB®

Virginia L. Stonick
Carnegie Mellon University

Kevin Bradley
Carnegie Mellon University

PWS Publishing Company

I(T)P An International Thomson Publishing Company

Boston • Albany • Bonn • Cincinnati • Detroit • London • Madrid • Melbourne • Mexico City
New York • Paris • San Francisco • Singapore • Tokyo • Toronto • Washington

PWS PUBLISHING COMPANY

20 Park Plaza, Boston, MA 02116-4324

MATLAB and PC MATLAB are trademarks of The Mathworks, Inc. The MathWorks, Inc. is the developer of MATLAB, the high-
performance computational software introduced in this book. For further information on MATLAB and other MathWorks
products--including SIMULINK™ and MATLAB Application Toolboxes for math and analysis, control system design, system
identification, and other disciplines--contact The MathWorks at 24 Prim Park Way, Natick, MA 01760 (phone: 508-653-1415;
fax:508-653-2997; email: info@mathworks.com). You can also sign up to receive the MathWorks quarterly newsletter and register
for the user group.
Macintosh is a trademart of Apple Computer, Inc.
MS-DOS is a trademark of Microsoft Corporation.
BookWare Companion Series is a trademark of PWS Publishing Company.

I(T)P™
International Thomson Publishing
The trademark ITP is used under license

For more information, contact:

PWS Publishing Co.
20 Park Plaza
Boston, MA 02116

International Thomson Editores
Campos Eliseos 385, Piso 7
Col. Polanco
11560 Mexico C.F., Mexico

International Thomson Publishing Europe
Berkshire House I68-I73
High Holborn
Loncon WC1V 7AA
England

International Thomson Publishing GmbH
Konigswinterer Strasse 418
53227 Bonn, Germany

International Thomson Publishing Asia
221 Henderson Road
#05-10 Henderson Building
Singapore 0315

Thomas Nelson Australia
102 Dodds Street
South Melbourne, 3205
Victoria, Australia

International Thomson Publishing Japan
Hirakawacho Kyyowa Building, 31
2-2-1 Hirakawacho
Chiyoda-ku, Tokyo 102
Japan

Nelson Canada
1120 Birchmount Road
Scarborough, Ontario
Canada M1K 5G4

About the Cover: The BookWare Companion Series cover was reated on a Macintosh Quadra 700, using Aldus FreeHand and Quark XPress. The surace plot on the cover, provided courtesy of The MathWorks, Inc., Natick, MA, was created with MATLAB® and was inserted on the cover mockup with a HP ScanJet IIP Scanner. It represents a survace created by assigning the values of different functions to specific matrix elements.

Editor: Bill Barter
Assistant Editor: Ken Morton
Manufacturing Coordinator: Ellen Glisker
Cover Designer: Stuart Paterson, Image House, Inc.
Editorial Assistant: Lai Wong
Marketing Manager: Nathan Wilbur
Production: Pamela Rockwell
Cover Printer: Henry N. Sawyer, Inc.
Text Printer and Binder: Malloy Lithographing

Printed and bound in the United States of America.
00 01 02 03 04 – 10 9 8 7 6 5 4 3

Contents

6. Speech Modeling, Prediction, and Synthesis 57

7. Image Deblurring and Decoding 69

Foreword

This laboratory text is intended to serve as a companion to a basic textbook covering signals and systems theory. To accommodate different instructors' interests and preferences, we have made the laboratory assignment ordering flexible, organizing it by application rather than by concept. We believe that learning signals and systems theory is best motivated by applications, and thus have elected to explore a variety of applications in these assignments rather than focusing on a single domain. Each application is introduced at a level requiring minimal theoretical background and can be revisited later as more advanced topics are covered. Background sections accompanying each assignment highlight the material needed to complete it.

To facilitate experimentation with applications as early in the course as possible, we treat fundamental signals and systems representations as a language. By familiarizing students with these standard representations early, they can confidently use MATLAB to explore and discover implications of signals and systems concepts in real applications.

The use of MATLAB as a computational tool is intentional. It allows so many theoretical concepts to be easily implemented that it would almost be a crime not to use it in this field. MATLAB raises programming to a data-processing abstraction, thus minimizing the dependence on strong coding abilities to get results. We assume that while students have had some exposure to computers and programming, they will not have any familiarity with MATLAB. Hence, we begin with a tutorial covering the basic concepts, tools, and syntax needed for problem solving using MATLAB. The use of MATLAB does not, however, remove our emphasis on practicality. By including exercises on topics such as finite-precision effects and unmodeled nonlinear dynamics, we attempt to introduce beginning students to some of the many differences that exist between theory and practice, as well as some of tricks and pitfalls in bridging this gap.

The early labs focus primarily on processing audio signals, which are familiar to students and allow them to develop an intuitive feel for their results. This necessitates a basic hardware platform, namely a PC with audio capabilities. Any computing environment that can run MATLAB and support audio playback and recording is satisfactory. Some labs require extensive computing power, such as the image processing assignment; we recommend that a more powerful platform be used for these labs so as to eliminate long operation times.

To support general computing environments, we supply a utility that can convert from almost any popular audio format (.wav, .voc, .au, .aiff, .sf, to name a few) into MATLAB files and vice versa. While the latest version of MATLAB (4.2 at the time of this writing) has support for some of these, this utility can support a wider variety and removes some restrictions on the hardware support necessary. The interface to this utility is provided through MATLAB functions that set the appropriate switches and make the program calls.

These labs have been developed at Carnegie Mellon University for use in a first course in signals and systems that serves as an introduction to courses in communications, controls, and signal processing. We are planning to use some of the more advanced sections of laboratory assignments in an introductory DSP course. We cover applications from all of these different areas in the assignments in an attempt to stimulate the student's interest in the possibilities of a subject area as well as a foundation for further study. We hope that you will find them both interesting and challenging and worth the doing!

Acknowledgments

The development of this laboratory book has been fun, exciting, and enlightening—in addition to a lot of work. It could not have been completed without the support and encouragement provided by a large number of students, faculty and administrative colleagues, friends and family, editors, reviewers and support staff. We cannot possibly cover all who had impact but will attempt to acknowledge significant contributions by individuals.

This project was made possible as a result of the new integrated Electrical and Computer Engineering degree and curriculum developed at Carnegie Mellon University, designed to be application-oriented. In the process of redesigning the first signals and systems course around a "just-in-time-learning" laboratory experience, we developed the physical laboratory and designed the basic laboratory assignments contained herein.

We worked with a number of graduate and undergraduate students in creating these assignments as part of either undergraduate projects or on-going graduate research projects. By working with students in developing the laboratory experiments, we were much more likely to catch potential points of confusion. Knowing that fellow students participated in developing the assignments is empowering; students persevere as they believe both that they have the ability to complete the assignment, and that it will prove to be interesting.

We would like to acknowledge the following individuals' contributions to the laboratory and assignments: G. Woyansky and F. Coetzee, for hardware, software, and troubleshooting of lab assignments and demos; D. Massie of E-mu Systems, E. Sabol, A. Schultz, and V. Mazzotta for the pitch shifting and digital filter design ideas; U. Pillai and S. Schweizer, for the speech modeling and synthesis lab; D. Brown, for work on the image processing assignment; B. Burke, for testing and debugging the analog filter design and AM assignments; J. Stephan, for the Ball and Beam idea and technical support; P. Renucci, H. Metla, and A. Nigam for refinements to the Ball and Beam tutorial; and all of the undergraduate and graduate students with whom we have worked for contributions to the ideas and technical design of these assignments.

We also would like to acknowledge Professor Steve Director, Dean of CIT and Whittaker Professor of ECE, for encouraging creative course development; Professors White, Stancil, Kumar, and Hoburg for ensuring availability of laboratory resources; Dr. Susan Ambrose, Director of the University Teaching Center, for sharing valuable experience in course planning and organization; Tom Robbins, for his faith in this project; Dr. John Cozzens, for wine and fishin' philosophy; Dana Massie, for reminders that life and work can be fun; Professor Tom Alexander, for ever present reminders of the joy in discovery; Dr. Molly Johnson, for evolving explorations of the meaning in life, work, and friendship; Dr. John Stonick, for life sustenance; and our families, for providing a sense of balance and support. This project could not have been attempted without the National Science Foundation Presidential Young Investigators Program, funding from which enabled more time to be spent on integrating research with education.

Many thanks also to our reviewers, for their thoughtful commentary and inciteful insight: Erik Cheever of Swarthmore College, Can Isik of Syracuse University, and Colin Wight of New Mexico Tech.

And thanks to S.P., for providing cookies.

For the Student

Most textbooks have an order of operations imposed on them. Topics are covered in lecture and readings in the text, and then you do the homework problems at the end of the chapter. This laboratory textbook is not entirely like that.

Instead of chapters covering a topic or two, each experiment covers several different topics that are all used in a given application context. We have experiments on speech synthesis, filter design, digital communications, controlling the placement of a ball on a beam, and many others. Each "chapter" in our book is an application of signals and systems concepts and practices, and should provide you with some physical insight into the mathematics presented in your text and lectures.

These experiments are based on the assumption that linear system analysis can be applied to a number of different areas, like control systems, communication systems, and signal analysis. You are not necessarily an expert in or may not even have heard of any or all of these areas, so each experiment is probably an introduction to something unfamiliar and potentially exciting. We believe that some basic signals and systems techniques and representations —e.g. frequency transforms, signal construction, differential/difference equations—can be used as basic descriptors of these systems; more complex areas can be reduced to a few of these basic operations so that they can be understood at an intuitive level. Don't be afraid if you don't initially entirely understand the problem or what should be done—in the real world, problems are very rarely well-defined or understood. The goal is to increase your level of understanding and awareness of the concepts and application by the time you complete the assignment.

You should read the background section and be familiar with the topics discussed in it before attempting to do the Lab Prep problems. The Lab Prep problems are physically positioned prior to the background section in each assignment so that you can focus your reading on the fundamental ideas needed to answer these questions. These problems will help you in two ways: you will have some practice in solving problems in the area of interest, and you will have a start on the experiment to be run during lab. See page xv for details on the chapter layout.

You may encounter implementation problems during lab. Things like "my function doesn't work—I get a matrix size multiplication error" to "I can't play sound anymore on my system" are likely to occur. Overcoming some of these will simply require you to check the validity of your implicit assumptions, e.g., you may have passed a column vector to a function when a row vector is required. Some difficulties will be artifacts of your computing environment that can be discovered and dealt with through experience and with the help of your instructor. Some problems illustrate standard difficulties encountered in practice and standard solutions; problems arise as the mathematics is not a completely accurate description of the application. In these cases, there are no "right" answers, just trade-offs that need to be explored in the specific application context.

We hope that you learn something from these labs, and we especially hope that you gain an appreciation for signal processing and linear system analysis. After all, that's the point.

Using This Text in Your Course

We have found it useful to be aware of the educational philosophies guiding the textbooks that we use in our courses. We therefore would like to inform you of the goals for which we aimed, the tools provided by our laboratory format, and how these can be used in course design.

These labs have been used in a single-semester introductory course on signals and systems at Carnegie Mellon University, where the prerequisite is an introductory circuits course. We require some basic understanding of linear algebra and complex variables, which we do review during lecture. Laboratories related to building and designing analog circuits also require a basic understanding of circuits. We assume that students have had some programming experience, but not with MATLAB.

We—from our experience as researchers and educators in the field—believe that it is the understanding and interpretation of the mathematics involved that makes signals and systems interesting and exciting. In our labs we focus on bringing the mathematics together with an application at the lowest level of mathematical sophistication possible so as to achieve an intuitive feel for the theory. This means that many of these laboratories can be done in the time domain first, if desired, and can be revisited to enhance the solution and its understanding using Fourier, Laplace, or z-transforms. Many of the important concepts can be easily tied to the application in the time domain; the difficulty in the past has been the lack of a single tool that facilitated simulation of complex systems from a basic conceptual understanding. MATLAB provides us with such a tool.

With a basic understanding of MATLAB, students have a tool for independent discovery of more advanced topics and can experience the real-life trade-offs evident in all practical engineering problem solving. To create a foundation for and facilitate this discovery, we provide significant guidance for the basic laboratory problems—so that students feel empowered and confident to proceed—and then provide additional laboratory problems that are more open-ended and encourage students to experiment with ways to improve their solution. Since students can significantly differ in the educational methods that work best for them, we have tried to present the problem in many different ways within each laboratory assignment. Each assignment includes goals and specific learning objectives, a discussion of the application context, a review of the basic mathematics needed in the assignment, as well as specific statements of the individual problems and directive questions to help students focus their efforts in productive directions.

In any laboratory setting, preparation is critical for making consistent progress. We have provided Lab Prep problems that ask the students to answer some simple questions based on the laboratory background material. These questions typically fall into two categories: simple examples to be worked out by hand, illustrating the basic idea that will be implemented in an application context in the laboratory, and formulation of the problem information for use with MATLAB, including computation of initial design parameters. The first type of question helps students to evaluate whether or not the results generated by their MATLAB experiments make sense and gives them an understanding of relevant "back-of-the-envelope" calculation. Such calculations are the key to any practicing engineer's success in evaluating the potential of design alternatives with minimal effort and cost. The second type of question is intended to get students to prepare in advance for the actual lab work. This skill is valuable in engineering practice, as lab time is typically shared and costly. For the more difficult Lab Prep problems, we often have students collaborate in groups to solve the Lab Prep problems during a recitation section prior to the laboratory period. We have found that this approach, integrating group problem solving with laboratory assignments, helps to enhance the laboratory learning experience by allowing students to focus on the theoretical aspects prior to coming to lab, where they often become immersed in implementation details.

Laboratory Assignment Format

With the exception of a few introductory laboratories, each laboratory assignment has the sections described below. We have organized the laboratories in this way in an attempt to maximize flexibility for different course organizations.

Purpose: Describes the overall goal of the laboratory in the application context. Typically we ask students to reread this statement of purpose after completing the laboratory to evaluate if this goal was achieved and to include this evaluation in their lab report.

Objectives: Describes specific functional tasks that students should be able to perform by the end of the laboratory assignment. Typically these are not expressed in terms of the specific application, but rather as general intellectual competencies and laboratory skills. Again, we have students evaluate and report on their level of achieving these objectives.

References: Provides a list of topical references to concepts addressed in the lab, with additional bibliographic references for applications, if needed. Topical references are separated into Review and Exploratory topics. Time or frequency domain review topics should have been covered in other coursework prior to attempting time or frequency domain problems, respectively, in the assignment. Exploratory topics are concepts explored in the assignment, and this list provides pointers to additional information that can be found in most introductory textbooks for signal processing, communication, and control systems.

Lab Prep: Problems to be completed before the laboratory session. As described above, attempting to answer these questions enables students to clarify any questions they have and to make more effective use of laboratory time. We have found that physically placing these problems prior to the background reading is very effective for helping students to focus their reading and extract critical information from it.

Background: This section provides a brief overview of the application scenario and concepts needed in the laboratory assignment. The review of concepts is focused on the application context, in an attempt to bring students to the point where they should be able to begin the assignment, and is not meant as a substitute for classroom or textbook instruction.

Laboratory Exercises: Statements of problems to be solved in the laboratory. For each problem, there is an in-depth problem statement and directive questions to help students focus on what they are trying to achieve. With most laboratory assignments, there are three types of exercises from which the instructor may customize the laboratory assignment:

1. *Time Domain questions*: These questions can be explored and answered in the time domain.Using these, beginning students can acquire some physical intuition and mathematical understanding of physical signals and systems. Only minimal familiarity with some basic mathematical representations of signals is assumed

2. *Transform Domain questions*: These are questions to be explored using Laplace, Fourier, or *z*-transforms. Typically these are for more advanced students or study. From these, students learn about the power of transforms for designing and understanding the effect of systems on signals.

3. *Discovery questions*: These questions are more open-ended and may use either time or transform domain concepts, or both.These questions provide direction for exploring further properties of the application or tools developed in the assignment and are more reminiscent of design problems. From these, students discover more about the trade-offs inherent in analyzing and synthesizing physical signals and systems.

Laboratory and Coursework Integration

Coordinating a laboratory with a course, to ensure that laboratories have maximal benefit, is a challenging exercise. Students need to have enough knowledge to begin and complete laboratory assignments but also need to have lab assignments not lag so far behind the class topics that either more interesting or advanced topics—that would benefit from reinforcement with lab instruction—are not addressed. In our experience developing these laboratories, we have identified four elements of course organization and administration that were critical for success.

1. *Introduce the idea of sampling within the first week of class.* This is important, as MATLAB exercises require that signals be in discrete time, while the physical signals we use are inherently continuous time. We accomplish this objective by discussing sampling in the time domain, defining sampling as a simple conversion from CT to DT signals by "picking points off the graph", and defining playback as interpolation to convert from DT to CT signals.

2. *Focus more on the representations used in signals and systems,* and how these are used in MATLAB, rather than focusing on the development of properties. Although MATLAB is an intuitive, powerful, and easy-to-use tool for professionals well versed in the standard signals and systems representations, it is not at all simple for first-time students. Once students are adept at putting information in the form required for exploiting powerful MATLAB tools, properties and performance can be easily discovered and explored in the laboratory.

3. *Make Lab Prep problems worth some credit.* Early in the semester, students are much more motivated to do Lab Prep problems prior to lab time if they get some course credit. We (and the students) learned that doing the lab prep is key to effectively using lab time and expediently completing the labs. We require that all students make a reasonable attempt at Lab Prep problems before coming to lab in order

to be able to receive full credit on the corrected Lab Prep problems in their report.

4. *Provide explicit instructions for laboratory report format.* Students need guidance in how to provide clear, concise, and accurate accounts of their laboratory explorations. In Appendix C is a copy of the laboratory write-up instructions that we provide for our students. We found that providing very detailed feedback for the first report substantially enhanced student performance on later reports.

5. *Provide "framing" at the beginning of laboratory periods.* We found that having laboratory instructors summarize the major goals, theory, and procedures for the assignment at the beginning of the laboratory period helped the students to better proceed with the lab and understand its implications. A similar summary to frame theory in terms of the application at the beginning of laboratory-related recitation periods also was found to enhance student focus and comprehension.

We found that if we adhered to these four guidelines, students felt confident and were prepared for and enthusiastic about exploring signals and systems concepts in laboratory applications. As a result of this laboratory at CMU, students have an increased understanding of the relationships between physical systems and mathematics and are more interested in pursuing advanced topics in signals and systems.

Laboratory and Course Schedule Organization

What follows is a 15-week semester schedule to which we have converged after several semesters of teaching this course, assuming two in-class exams per semester occurring in weeks 6 and 11. Lab Numbers correspond to assignment numbers in this text. Since there is some overlap in laboratory assignments, we tend to rotate in different assignments each semester.

Table 1: Suggested Course of Study

Week	*Class Topics*	*Lab*
1	Plotting complex-valued functions, CT and DT signals, sampling	1. MATLAB Introduction
2	Operations on signals, systems as operators.	2. Audio Lab Introduction
3	Fourier Series and Transform as applied to signal spectra	3. Music Synthesis (time domain)
4	System properties, CT system representation as differential equations: I/O and state space, differential operator and transfer function.	3. Music Synthesis (frequency domain)

Table 1: Suggested Course of Study

Week	Class Topics	Lab
5	DT systems as difference equations: I/O and state space, delay operator representation of transfer function	5. Prosthetic Limb simulation
6	DT unit pulse response and convolution, convolution examples	6. Speech Synthesis
7	CT unit pulse response and convolution, convolution examples	Make-up
8	LTI frequency response	7. Image Processing (deblur)
9	CT response to e^{st}, Laplace transforms	7. Image Processing (decode)
10	Laplace solution of differential equations	8. Digital Communications
11	System realizations, poles, zeroes, and system stability	11. Ball and beam control problem
12	Analog filters and pole/zero locations	Make-up
13	Digital filters and pole/zero locations	9. Filter design
14	Fourier transform and properties	12. Amplitude Modulation
15	AM and sampling	Free

This arrangement of topics may seem to go against the grain of traditional signals and systems instruction; frequency domain representations are introduced early in the semester and discrete time topics are intermingled with continuous time. Also, some lab topics, like digital sonar localization and signalling may seem to be too advanced. We try not to shroud things like root locus and state space representations in a cloud of mystery; these are fundamental tools for solving problems and are not really very hard concepts to master. It is important to note that fundamental concepts, e.g., sampling, convolution, transfer functions, and filtering, can readily be applied to seemingly advanced topics, exposing the students to a wide variety of signal processing and control areas while not limiting them to trivial experiments designed to point out particular aspects of theory. It is possible, after all, to do something without full theoretical understanding of the problem, which develops best from hands-on experience.

The laboratory assignments in this text are flexible enough to easily accommodate more standard two-course sequence organizations. Using the Review and Exploratory topic lists as a guide, and applications that most intrigue them, instructors can design the laboratory segment of a course to be an exciting experience for both faculty and students. We encourage you to experiment with nontraditional course organizations in an effort to determine how to best generate enthusiasm and enhance motivation for each new crop of students.

Laboratory Assignment 1

MATLAB Introduction

PURPOSE

In this lab, you will be introduced to MATLAB by using it to evaluate expressions containing complex numbers. Using MATLAB, you can easily express these complex numbers in either rectangular or polar form. MATLAB's plotting capabilities will be introduced and used extensively. You will also use M-files and create some simple signal processing functions that are used in later laboratory assignments.

1.1 OBJECTIVES

By the end of this laboratory assignment, you should be able to:

1. Use MATLAB to perform complex arithmetic.

2. Generate and plot signals and complex valued functions.

3. Confidently develop MATLAB M-files and save results of computations from a MATLAB session.

1.2 REFERENCE

Review Topics

1. Algebra of complex numbers

2. Sketching of discrete and continuous time signals

3. Vector and matrix algebra

1.3 INTRODUCTION

As you work through the problems and experiments in this text, you will be using MATLAB—a powerful computing environment for numeric computation and visualization—quite frequently. MATLAB is designed for ease of use and behaves as a high-level programming language that is tailored for signal processing, communication, and control tasks. It is used by professionals in industry and academia worldwide in research, development, and design.

MATLAB, short for MATrix LABoratory, works on matrices of numbers. We focus mostly on one-dimensional matrices called vectors that contain signal samples, or on multiple-dimensional matrices containing several signals or the parameters of a system. For example, a vector could contain just a list of values from a mathematical function that you wish to plot. We will first focus on familiarizing you with the matrix notation in MATLAB, and get you used to working with vectors and matrices in arithmetic operations.

Everyone has their own view of how best to learn MATLAB. Our view is that things are best learned by first reading, then doing. This introduction has three parts: reading, reading while doing, and doing guided by knowing expected results. Since the main difficulties in learning MATLAB are in learning the syntax and (in some cases) learning to program, we hope that this format will remove some of the trauma. Experienced MATLAB users should try the Quiz to be sure they remember everything.

This introduction is not intended to present you with everything you need to know about MATLAB; it is merely to bring you to a point where you can do the labs contained in this book. Use the manuals (probably available from your computing systems personnel), on-line help, and references for additional information.

You should follow this text with MATLAB running, and work through the examples and questions.

Using MATLAB

MATLAB is available on Macintoshes, IBM PCs, DEC workstations, Sun workstations, and many other platforms. MATLAB runs under Microsoft Windows, X Windows, or in the Macintosh window environment.

All versions of MATLAB are compatible in file storage format and M-file format, so data stored on one system can be transferred to another without loss. Each MATLAB session has at least two windows: a text window, where commands are typed and data is displayed, and a graphics window, where graphics appear.

Starting MATLAB:

1. IBM MATLAB: after starting Windows, find the MATLAB icon and double click. Two windows will open.
2. X Windows MATLAB: after starting X Windows, type **matlab** at a shell prompt. If the program is on your path, it will start. The shell window becomes the text window.

MATLAB has on-line help for all functions and a set of demos; we recommend that you try the demos so you can see some of MATLAB's computing power and some of the functions available to you.

Tutorial, Part I

The following sections highlight some useful MATLAB commands by working through some example problems. You should work along with the text.

1. Starting MATLAB

   Start up MATLAB on the system you choose to use to do this assignment. When the text window opens, a prompt appears:

   >>

   All commands will be entered after a prompt like this one.

2. Evaluating Complex Variables and Expressions

   In the following text, information that you enter will be preceded by the MATLAB prompt; information not preceded by a prompt is printed by MATLAB as a computation result or other information.

Problem 1.   Express each of the following complex numbers in Cartesian form, i.e., $s = a+jb$, where $a = \text{Re}\{s\}$ and $b = \text{Im}\{s\}$. Plot part (a) in the complex plane.

   a. $je^{j11\pi/4}$

   b. $(1-j)^{10}$

Part a: You can find the Cartesian form by typing the expression, using standard symbols for arithmetic operations.

```
>> j*exp(j*11*pi/4)
ans =
-0.7071 - 0.7071i
```

Note that MATLAB has evaluated the expression and echoed the result to the screen, expressed in Cartesian form as the variable **ans**. Also, just like any programming language, **exp(x)** returns e^x. Other standard functions, including trigonometric functions, are available; type **help elfun** for a list. Additionally, **pi** is defined as a special variable having the value π, and **j** is defined as $\sqrt{-1}$. Any special variable will act as defined until you change its value by assigning a new value to it. For example, to change **pi** to 3, issue the command

```
>> pi = 3
pi =
        3
>> pi
pi =
        3
```

As you can see, **pi** has been changed. If you clear your definition, the old one reappears:

```
>> clear pi
```

```
>> pi
ans =
        3.1416
```

Other variables can be set in the same way:

```
>> z = 3+4*j
z =
        3.0000 + 4.0000i
```

So, to set z to the solution of part b:

```
>> z = (1-j)^10
z =
        0.0000 -32.0000i
```

3. Plotting Complex-Valued Functions

Plots in MATLAB are generated using the plot function.

---

plot(x,y) generates a plot where the values of the vector **x** indicate points along the horizontal axis corresponding to the values in the vector **y** that are to be plotted on the vertical axis. Vectors x and y must have the same number of elements.

Since complex values have two components corresponding to **a+jb**, MATLAB provides the **real** and **imag** functions to separate the real and imaginary parts of an imaginary number:

```
>> z = 3+4*j;
>> zr = real(z);
>> zi = imag(z);
```

real and **imag** break **z** into real and imaginary parts in the variables **zr** and **zi**, respectively.

Note that typing a semicolon at the end of the command line prevents MATLAB from echoing the result back to your screen. This will be important when you create large matrices and vectors. In this example, the value of **z** was not printed as it was above.

To plot a complex number, we can either plot the real parts vs. the imaginary parts or let MATLAB do it for us. Here, we supply **zr** and **zi** to the plot function:

```
>> plot(zr,zi,'x')
```

and MATLAB generates a plot:

Laboratory Assignment 1

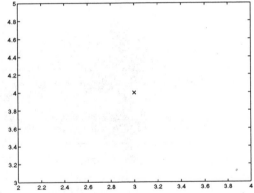

Figure 1.3.1 *Example plot of a complex number*

> The `'x'` parameter to the plot function tells MATLAB to generate an x shape for each data point instead of a 'connected-dot' display. Since we only plotted one data point, this is extremely useful. In general you should always label axes on your plot and include a title. **help plot** shows you the other characters that can be used as well as the different colors that can be used on the plot.

This plot would be exactly the same if we had entered:

```
>> plot(z,'x')
```

since the MATLAB default for plotting complex numbers is to plot the real parts on the horizontal axis and the imaginary parts on the vertical axis.

Multiple sets of parameters can be given to **plot**; each pair of arguments is taken as x and y data pairs.

If you wish to have several plots shown at once on different sets of axes, use **subplot**—see the on-line help—or open new figures by typing **figure** and creating a whole new plotting window.

Vectors and Matrices: Tutorial, Part II

Matrices and vectors make up the heart of MATLAB computations. In this section, matrix and vector manipulations will be introduced. A vector is a one-dimensional list of values, an m x 1 or 1 x m matrix. Vectors hold single signals or lists of data. They can be assigned a name and treated as any other variable in MATLAB; however, operations performed on vectors are done element by element.

As an example of this, consider the function $y = 3x+2$. If we want to plot y as a function of x, we first create an x vector containing data points in the range of interest. Suppose the range is 0 to 5, using every integer point. There are several ways to create this data set in MATLAB. The first way is to type in every point:

```
>> x = [ 0 1 2 3 4 5 ];
```

This generates a row vector **x**, i.e. a 1 x 6 matrix containing six elements, the integers 0 through 5. Note that the semicolon keeps MATLAB from echoing the results of your command back to the screen. An easier way to generate this same vector is to use a range-generating statement:

```
>> x=0:5
x =
     0  1  2  3  4  5
```

The colon operator acts like the word "to", in effect generating the function "0 to 5". A step size of 1 is the default. A different step size, positive, negative, real or integer, can be specified by placing the step value between the beginning and end of the range, as in **z** below:

```
>> z = 0:0.01:5;
```

generates 501 data points that are 0.01 apart, starting from 0 and ending at 5.

The next step is to evaluate the function **y**, using the **x** as defined above:

```
>> y=3*x+2
y =
     2  5  8  11  14  17
```

This statement instructs MATLAB to multiply every element in **x** by 3 and then add 2 to every element, storing the results in **y**. Thus 3***x** is treated as scalar multiplication of a vector and the 2 is implicitly treated as a vector of the same length as **x** comprising all 2s.

Since MATLAB is based on matrix operations, it is important to recall that you can only add or subtract matrices having the same dimensions, e.g., the addition of a **3x2** matrix with a **2x3** matrix is undefined. Matrix multiplication requires that the number of columns in the first matrix be the same as the number of rows in the second matrix. For example, multiplication of a **2x5** matrix **A** with a **5x3** matrix **B** results in a **2x3** matrix **C=AB**, whereas the multiplication **BA** is undefined. However, the multiplication **D=B'A** is defined, where **'** denotes the transpose operation in MATLAB.

---

1. Generating Complex Functions

Let's generate values for the complex function $f(t) = 3e^{j3\pi t}$ for t ranging from 0 to 1 in 0.001 increments. The first step is to create a time variable; note the use of the **:** operator with a non-integer step size.

```
>> t = 0:0.001:1;
```

Here the semicolon at the end is especially important, unless you really want to see and wait for all 1000 values to be echoed back on your screen.

Laboratory Assignment 1

Next, construct a vector containing values of this function for each time value in **t**:

```
>> f = 3*exp(j*3*pi*t);
```

The variable **f** now has the complex result of the function evaluation.

> It should be pointed out that transcendental functions (e.g. **sin**, **cos**, **exp**) in MATLAB work on a point-by-point basis; in the above command, the function **exp** computes a vector where each element is the exponential of its corresponding element in **j*3*pi*t** (1001 total elements).

2. Accessing Vectors and Matrices

The data in vectors can be viewed and displayed in several different ways: it can be plotted, printed to the screen, printed on paper, and saved electronically. It is not, however, always desirable to access the entire vector at once when displaying the information in it. To access single elements or ranges of a vector, an index element or list that identifies which elements are of interest is needed.

Elements in MATLAB vectors are identified by the vector name and an integer number or index, much in the same way that DT signals are indexed by integer values. However, in MATLAB, only positive integer indices are used. Thus the first element in a row or column vector **f** is denoted by **f(1)**, the second element by **f(2)**, and so forth. To access specific elements in a vector, you need to use the name of the variable and the integer index numbers of the elements you wish to access. Range statements can be used for indices to access the indexed elements much in the same way that range statements are used to define vectors comprising values in a specified range. For example,

```
>> f(25);
>> f(3:10);
>> f(1:2:50);
```

The first line accesses the 25th element of **f**. The second accesses elements 3 through 10, inclusive; and the third statement returns the odd-numbered elements between 1 and 50.

Elements in matrices require use of two-dimensional indices for identification and access. For example, **f(3,2)** returns the element in the third row, second column; ranges can also be used for any index. For example, **f(1:3, 4:8)** defines a matrix that is equivalent to a section of the matrix **f** containing the first, second, and third rows, and the fourth through eighth columns.

> If a **:** is used by itself, it refers to the entire range of that index. For example, a 3 x 5 matrix could have its fifth column referenced by **f(:,5)**, which means "all rows, 5th column only," as well as **f(1:3,5)**, which means rows 1 to 3, 5th column only.

The index number can be another variable as well. This is useful for creating programming loops that execute the same operations on the elements of a matrix.

From other computer programming experiences, you should be familiar with the idea of creating a loop to repeat the same task for different values. Here's an example of how to do this using MATLAB. Suppose we want to generate an output vector where each element is the sum of the current element and the element from 10 back in an input vector. The task to be repeated is the sum of two elements; we need to repeat this for each element in the vector past 10. The elements of the vector **x** define the input ramp function to be integrated, **y** will hold the result, and **k** is the loop index:

```
>>x = 3*(0:.1:5) + 2;
>>y=zeros(size(x));
>>for k=11:length(x),
y(k)=x(k-10) + x(k);
end
>>
```

When a loop structure is entered in MATLAB, the body of the loop does not have a prompt printed; the command line acts the same as with a prompt. *Note that the loop will not be executed until the* **end** *command is entered and followed by a carriage return.* You will wait forever for the loop results if you leave off **end** (indicating to MATLAB that you have done all you wanted to in the loop).

As you can see, the variable **k** is set to range from 11 to the length of **x**; this allows **k** to index all elements in **x**. **k** is set to increment by integers.

Loops generally are not desirable, since they take a very long time to run. If possible, rewrite your operations in terms of vector additions and multiplications instead of looping.

For example, we can rewrite this problem to use vector addition by creating two new vectors, one which is **x** offset by 10 and the other which is **x** padded with 10 zeros (since we can only add vectors of similar lengths).

```
>>x = 3*(0:.1:5) + 2;
>>x1 = [zeros(1,10) x];
>>x2 = [x zeros(1,10)];
>>y = x1+x2;
```

Note that using vector addition in MATLAB is significantly more efficient than using **for** loops.

Storing Results and M-Files

Usually you want to save the results that you have generated during a MATLAB session, including data vectors created and commands used to process them. This can be accomplished by

---

1. Using the **diary** command to save a record of all commands typed. Enter **help diary** to learn how to use this command.

2. Using M-files that you have created using a text editor; such files have a **.m** extension and contain a list of MATLAB commands to be executed when you type the root filename.

3. Saving the contents in variables for the next session using the **save** command.

It is recommended that you become familiar with M-files. They are extremely useful and will save you much time and effort. There are two types of M-files: scripts, which are essentially a series of commands typed into a file instead of typed at a MATLAB prompt; and functions, which allow you to create new MATLAB functions.

Some of the later problems will use M-files as part of the solution. To edit M-files, any text editor can be used—not word processors like MS Word or Framemaker. The Macintosh versions of MATLAB have a built-in editor; PC and X Windows versions rely on outside editors such as emacs. Note that these editors can be run in a separate window to allow editing and processing to occur concurrently.

1.4 ASSIGNMENT PART I

Now that you have been introduced to MATLAB, larger and more complex problems can be handled. This section will present an example of how to solve a problem using MATLAB. In addition, you will be given several problems with answers, but not the steps taken to generate the answers. You should work through these problems during the laboratory session.

Example: Functions of Vectors

Question 1.   Add the functions $p=3\sin(x^2) + 2\cos(y^3)$ and $q=3\cos(xy) + 2y^2$ for x in the range 0 to 5. $y=0.05x+2.01$ in all cases. Use increments of 0.01. Plot all three functions.

Solution:

This can be done in two different ways. The easiest way is to make two vectors containing the individual functions and a third vector containing the sum. The alternative is to algebraically substitute in x, y, p and q into the total expression and reduce. This is extremely unpleasant.

> Note that when a dot precedes an operator, as using **.*** for multiplication, it implies that each element in the vector (matrix) results from applying that operator to corresponding elements in the first and second vectors (matrices). For example, dot multiplication of two m x 1 vectors results in an m x 1 vector where each element is the product of the corresponding elements in the first and second vectors. Note that **.*** is equivalent to the "dot-product", or inner product operation from linear algebra. This type of multiplication requires that the vectors (matrices) must be of the same size and is called pointwise, rather than vector or matrix, multiplication.

```
>> x = 0:0.01:5;
>> y = .05*x+2.01;
>> p = 3*sin(x.*x)+2*cos(y.*y.*y);
>> q = 3*cos(x.*y) + 2*y.*y;
>> z = p + q;
>> plot(p);
>> plot(q);
>> plot(z);
```

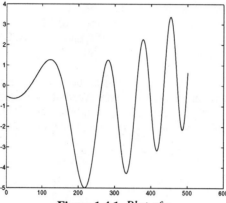

Figure 1.4.1 *Plot of p*

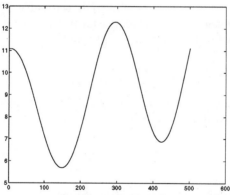

Figure 1.4.2 *Plot of q*

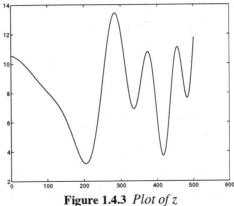

Figure 1.4.3 *Plot of z*

Question 2. Plot the function $y(t) = (1-e^{2.2t})\cos(60\pi t)$. Use t from 0 to 0.25 in 0.001 increments.

Answer:

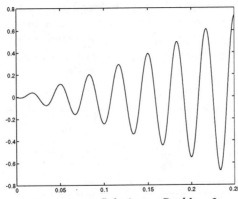

Figure 1.4.4 *Solution to Problem 2*

Question 3. A polynomial function has roots at -2, 2, -2+3j, -2-3j. Determine the polynomial, plot the four roots in the complex plane, and plot the polynomial function for the range $x \in (-5, 5)$ in steps of 0.01. You may wish to use **help** to look up the functions **poly**, **roots**, and **polyval**.

Answer: The polynomial is $x^4 + 4x^3 + 9x^2 - 16x - 52$.

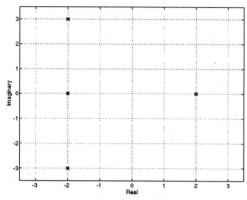
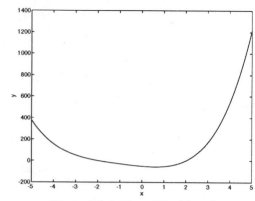

Figure 1.4.5 *Roots of Problem 3*

Figure 1.4.6 *Plot of Problem 3*

Question 4.   The complex function $f(t)$ has the form: $f(t) = 3e^{-j2\pi t + \pi/4}$.

Plot the real and imaginary parts as a function of time from 0 to 3 seconds in 0.01-second increments. Also plot the magnitude and phase of f as a function of time. You may wish to look up the functions **subplot**, **title**, and **plot** to see how to generate more than one plot in the graphics window at the same time. You will also need **abs** and **angle**. These commands are very useful in signals and systems.

Answer:

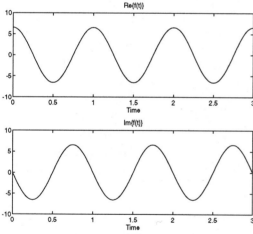
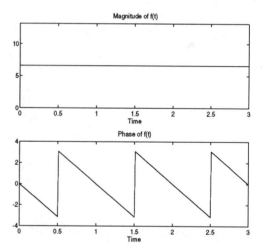

Figure 1.4.7 *Real and Imaginary plots*

Figure 1.4.8 *Magnitude and Phase plot*

This section will focus on the creation of MATLAB M-files. The solution to a problem is the M-file itself. We have provided you with the output of our M-file (without echoing the generation commands to the screen). This is done by using the command **echo off** at the beginning of the M-file.

Example: Creation of a Function in MATLAB

A function in MATLAB is a special kind of M-file. The first line in the file defines the function, both giving it a name and indicating what values are to be passed as arguments to the function and those that are to be generated by the function, much like a subroutine in other programming languages. Once you have created a function you can use it just as would use a MATLAB-supplied function.

> A useful feature of function files is that the lines that follow the function definition and begin with a comment symbol (**%**) are printed when help is requested for your function (**help yourfunction** prints these lines).

Make a function that takes two variables as arguments, adds 1 to the first variable, multiplies the second variable by two, and returns the product of the two variables. This function is called **blackbox**; the file is called **blackbox.m**

The following M-file is generated.

```
function [output] = blackbox(a,b)
% Adds 1 to argument a and multiplies b by 2
% Returns the product of a and b in the output.
%
% If the two variables are not of the same size,
% the larger variable is stripped to be the same
% size as the smaller.

% Determine the lengths of each vector
la = length(a);
lb = length(b);

% Add 1 to a
a = a+1;

% Multiply b by 2
b = b*2;

% Compare the lengths: if a is shorter than b, truncate b
% Otherwise, truncate a.
if la < lb
```

```
output = a .* b(1:la);
else
output = a(1:lb) .* b;
end
```

> The elements in this file that are required in order for MATLAB to recognize it as a valid function are (1) the filename and the function name must be the same, (2) the vector **[output]** is the function output, (3) and, finally, the M-file must start with the word **function** to indicate to MATLAB that it is a function that may require inputs.

Question 5.   Generate a file that calculates a sine wave of 5 Hz for 3 seconds using 0.001-second increments and plots the sine wave versus time with all axes labeled. Display the length of the time sample and the length of the sine wave calculated.

Answer: the script file sine. m

```
>> sine
Length of time vector
3001
Length of sine vector
3001
>>
```

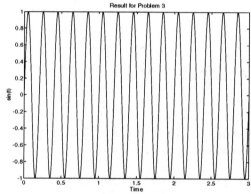

Figure 1.5.1 *Solution to Question 5*

Question 6.   Generate a function called squarer that returns the vector with each element squared. Use this function with the sine wave from the previous problem and display the sum of the elements in both the sine wave vector and the vector composing square of the sine wave.

Answer: the file prob5.m

```
>> prob5
Sum of sine
 9.5372e-14
```

Laboratory Assignment 1

```
Sum of sine squared
  1.5000e+03
>>
```

Note that the sum of a sine squared should be 0.5 if we interpret the sum as an approximation to an integral. The function returns 1500 because there are 1000 samples per period and three periods. 1500 / 3000 is 0.5. You must be cautious when interpreting results because you are dealing with discrete elements in the vector instead of a true continuous function.

Note also that if the sum of the sine function accurately approximates the integral, it should be 0. It is, instead, a very small number. This error is due to rounding errors in the least significant place of every sine value calculated and is an artifact of numerical computing. If you expect a sum to be equal to zero, don't be surprised if it is very small instead of zero.

A brief word about help: MATLAB's **help** can be useful, but it can also be confusing. If you type **help** by itself, you will see a list of directories and a description of their contents. **help <directoryname>** will show you the contents of a directory; **help <function>** gives you the help associated with the function. Sometimes these are cross-referenced, sometimes they aren't.

1.6 QUIZ

Problem 1. Compute $\dfrac{(2+j5)\ (1-j5)}{(1+j7)\ (3+j2)}$ and express your answer in both rectangular and polar coordinates.

Problem 2. An interesting plot is generated by the complex function $r = 1 - \cos\theta$, where r is the radius of a complex number z expressed in polar coordinates and θ is the angle. Sweep θ from 0 to 2π and plot the real and imaginary parts of $z = r\exp(j\theta)$ as x and y coordinates, respectively.

Problem 3. In communications, a signal is called a power signal if it has a zero time average $\bar{x} = \dfrac{1}{N}\sum_{k=0}^{N-1} x[k]$ and a nonzero, finite time average of its square $\overline{x^2} = \dfrac{1}{N}\sum_{k=0}^{N-1} x^2[k]$; in contrast, a message signal has a nonzero time average and a nonzero, finite time average of its square. Create a function that determines if a signal is a message signal or a power signal. It should return the value 1 if the signal is a message, and 0 if the signal is power. Test your function with the power signal $\sin(10\pi t)$ and the message signal random(t). Let $t = [0,1]$ and use increments in time of 0.01. Use the **randn** function to get a normal (i.e., Gaussian) distribution (see the on-line help for **randn**).

Problem 4. Some MATLAB functions require that a vector representing time samples be in the first column of a matrix and the corresponding signal values at each time be in the second column. Such a matrix, called *A*, has been created. Write down three MATLAB expressions to extract the time vector into the variable t, the signal vector into the variable x, and the 100th time/signal pair into the matrix y. *Hint: Use* **:** *notation.*

Problem 5.   Write a MATLAB function **half** that removes every other element from an arbitrary length vector, creating a shorter vector made of only the odd-numbered elements of the original vector; and a MATLAB function **double** that creates a longer vector by adding an additional element between neighboring elements in the original vector. Each new element should equal the average of its neighboring elements. Use only matrix/vector manipulations; do NOT use loops. Test your solution by applying **half** and then **double** to the vector **x = [1 2 ... 6 7 6 .... 2 1]**. What happens after the execution of **half** followed by **double**, and **double** followed by **half**?

Laboratory Assignment 2

Signal Sampling, Manipulation, and Playback

PURPOSE

This lab will introduce you to the laboratory equipment and the software that allows you to link your computer to the hardware. Specifically, you will learn how to use the equipment in the laboratory to convert analog signals (e.g., from a microphone or function generator) into digital signals that can be displayed and/or modified using MATLAB. You will learn how to perform mathematical modifications to digital audio signals in order to produce some interesting audio effects. In the process, you will learn how to take the original or modified digital signals and convert them to analog signals for display on an oscilloscope or for audio playback.

In this text, we refer to .wav files for convenience. Substitute whatever file type is necessary for your system, as discussed in the Foreword. Also, see Appendix A for a full description of the sound file conversion program.

2.1 OBJECTIVES

By the end of this assignment, you should be able to:

1. Use the laboratory setup to import real signals from a microphone or function generator into the computer and to export signals to an output device, such as headphones or an oscilloscope.

2. Use some simple mathematical manipulations to alter audio properties of signals.

2.2 REFERENCE

Review Topics

1. Mathematical representations of continuous- and discrete-time signals

2. Mathematical operations on signals

3. Representation of sinusoids at specified frequencies

1. Sampling, quantization, and A/D conversion

2. Digital audio signal manipulation using MATLAB

3. Laboratory audio platform capabilities and requirements

2.3 LABORATORY PREPARATION

Question 1.   In this lab, you will sample a 100-Hz tone at 8000 samples per second. If samples of this signal are stored as a row vector, what will be the effect of your MATLAB function file **half** on the pitch and duration of the tone if the playback rate is identical to the sampling rate? Similarly, what will be the effect of **double**? Be specific; indicate pitch in Hertz and duration in seconds relative to the original signal. Use sketches to help explain your answers. How can you use your functions **half** and **double** if the original sampled signal is stored as a column vector?

Question 2.   In this lab you will apply the MATLAB functions **fliplr** and **flipud** to a sampled segment of your speech. Use the MATLAB help function to find out what these commands do. For each function, explain the impact it will have on both row and column vectors. What will you expect to hear when you play back the signal returned by these functions, assuming that you use the original sampling rate and the signal is stored as a column vector?

Question 3.   Given a sampled signal contained in the vector **x** in MATLAB, write a MATLAB function to generate a vector that includes the original signal plus an echo occurring **T** seconds after the start of the original signal and having amplitude that is **a** times as large as that of the original signal. Use vector operations to avoid time-consuming loops.

2.4 BACKGROUND

In order to store, display, and modify audio signals on a digital computer, the physical analog signals must be digitized. This is done through two processes known as sampling and quantization, collectively referred to as analog-to-digital (A/D) conversion by convention. Here we provide a brief review of sampling using only time domain concepts; in a later chapter, we will revisit sampling and develop more analytical insight using frequency-domain concepts.

Sampling Review

Consider an analog signal $x(t)$ that can be viewed as a continuous function of time, as shown in Figure 2.4.1(a). We can represent this signal as a discrete-time signal by using values of $x(t)$ at intervals of nT_s to form $x[n]$, as shown in Figure 2.4.1(b). Essentially, we are "picking points off of the graph" of $x(t)$ at regular intervals. Mathematically, we write $x[n] = x(t)\big|_{t=nT_s}$, where T_s is called the *sampling period.*

Instead of specifying T_s, we can specify the sampling rate or frequency f_s (in Hertz) which is inversely related to the period by $T_s = 1/f_s$. Note that if T_s is 'small enough', then $x(t)$ can be reconstructed from $x[n]$ by simply "connecting the dots" (one of the simplest interpolation methods). Intuitively, the faster $x(t)$ varies, the smaller T_s must be to provide a reasonable approximation of $x(t)$. In order to ensure that the continuous-time (CT) signal can be exactly reconstructed from its discrete-time (DT) representation, the sampling rate must be greater than twice the highest frequency present in the signal; this rate is commonly referred to as the Nyquist rate.

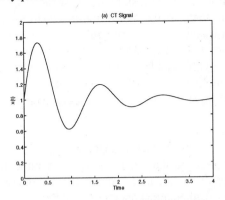 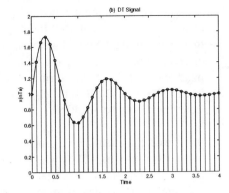

Figure 2.4.1 *A signal (a) and its samples (b)*

If the sampling rate is not sufficient, the samples of the signal will not accurately represent the signal. Consider sampling $x(t)$ in Figure 2.4.1 at only the positive peaks. If we connect the dots, the resulting CT function appears as a decaying exponential without oscillation; thus much of the information, or "character," conveyed by the original CT signal has been lost.

If the sampling rate is substantially higher than the minimum required, or Nyquist, rate, then the signal is called "oversampled." For example, some CD players oversample by a factor of 128 (a rate of 5.645 MHz) to enable low-cost, high-quality sound reproduction. Since MATLAB operations are inherently discrete time, and many physical signals are more naturally represented as continuous-time signals, the sampling rate is of considerable importance. Prior to sampling at a given rate, audio hardware typically eliminates high-frequency information that can't be represented at the given rate by using a so-called anti-aliasing filter. Standard sampling rates used in commercial technologies depend upon the expected characteristics of the signals used. Telephone-quality voice is sampled at 8 kHz, standard sounds in Windows 3.1 are sampled at 22 kHz, and CD-quality audio requires a 44.1 kHz sampling rate.

In general, if the data being processed is sampled less than twice per period, you need to increase the sampling rate. We usually specify the sampling rate so that your results will be accurate, but this might not always be the case in other circumstances.

Quantization

Quantization is the process by which signal sample values are represented as binary digits to permit manipulation on a digital computer. The number of bits used to define a sample indicates how many different analog levels can be represented. For example, using 8 bits per sample permits representation of up to 256 different levels, whereas 16-bit sampling can represent 65536 levels. The maximum difference between adjacent sampling levels determines how accurately the amplitude of the digitally represented samples approximates the sample values obtained from the original continuous-time signal. The more sampling levels there are, the more accurate the representation.

One commonly used measure of quantization accuracy is the signal-to-noise ratio (SNR) which is the ratio of the amount of power in the signal to that in the "noise" introduced by quantization errors. Larger values of SNR indicate higher-quality signals. For a full-range signal (meaning that representing the full range of signal values in the original signal requires using all possible discrete amplitude levels), the SNR can be roughly calculated as $(6n - 7.2)$ dB, where n is the number of bits. Using this formula, 16-bit sampling has a SNR of ~89 dB while 8-bit sampling has an SNR of ~41 dB. If the signal ranged from -1 V to +1 V, the quantization error for 16-bit sampling would be +/- 30 μV.

In essence, the quantization process introduces controlled errors so that we can use a computer to process the signals. Why would we deliberately introduce noise into a signal rather than simply using continuous-time, analog processing? There are many advantages, but the two major ones are that (1) typically, more sophisticated mathematical operations for signal processing, communication and control systems can be performed using a computer, and (2) it is much easier to reduce errors, caused by disturbances during transmission or degradation due to heat or aging, if the information is stored in a binary format: recovering the information requires only a choice between a binary one and zero, and codes can be used to help detect and correct errors. By introducing controlled error into the signal, we can prevent corruption from uncontrolled noise sources.

Loudness and Pitch in Audio Signals

Since you are likely to be using audio signals in several laboratory assignments, we would like to comment briefly at this point on how perceived loudness relates to signal amplitude, and how perceived changes in pitch relate to changes in frequency. These ideas will continue to be reinforced as you proceed through the laboratory assignments.

First, linear changes in perceived loudness correspond to exponential changes in signal amplitude since the loudness depends upon the signal power. For a simple tone (sinusoid) at 100 Hz, the average power over a single period is defined as

$$AveragePower = \frac{1}{T}\int_T \left(A\sin\left(\frac{2\pi}{T}t\right)\right)^2 dt = \frac{A^2}{2}$$

Loudness is usually measured in decibels, defined as $10 \log_{10}$(power). A signal that is twice as loud has twice as many decibels. Thus squaring the amplitude of the quieter tone will give you the amplitude of the tone that sounds twice as loud. (Verify this for yourself using tones and properties of logarithms.)

Pitch and frequency in audio signals also are logarithmically related. Doubling the frequency corresponds to increasing perceived pitch by one octave; increasing pitch by one note on the musical scale corresponds to increasing frequency by a factor of $2^{1/12}$. Specific correspondences between pitch and frequency are discussed in more depth in Laboratory 3.

A Mathematical Model of the Echo Process

In this lab you are asked to develop MATLAB code to generate an echo. A simple analytical model of an echo $s_e(t)$ of a signal $s(t)$ is $s_e(t) = \alpha(t\text{-}T)s(t\text{-}T)$, where the $\alpha(t)$ function attenuates the volume over time and T is the time delay introduced by reflection. In this treatment, we will assume that $\alpha(t) \in [0, 1)$ and $T \geq 0$. This implies that the echo's strength is weaker than the true signal and that the echo always arrives at a time later than the true signal (post-echo). The signal $r(t)$ that you hear is formed by a simple addition of the original signal and its echo. This process is illustrated in the figure below for the simple case where $\alpha(t)$ is a constant.

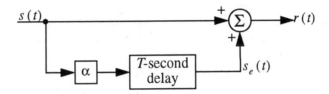

Figure 2.4.2 *Echo model*

In this figure, $s(t)$ is a transmitted signal, $s_e(t)$ is that signal's echo, and $r(t)$ is the signal received by a listener. Thus, in general, $r(t)=s(t)+s_e(t)$.

2.5 LABORATORY EXPERIMENT

For each problem below, make sure to note the specific signal parameters and equipment settings and connections that you used and the procedure followed to select values resulting in good performance. This will help you in writing your report as well as in later assignments.

Problem 1.   Here you will see how changes to frequency, amplitude, and waveshape of signals affect how they sound.

   1. Hook up the output of the function generator to the oscilloscope.
   2. Generate a sine wave having a MAXIMUM amplitude of 1 V.

3. Also connect the output from the function generator to the input (line in) of the sound card.

4. Connect the headphones to the appropriate sound card output.

5. Listen to the headphones and watch the oscilloscope as you vary the frequency, amplitude, and waveshape (square, triangular, sinusoidal). Notice how the audio properties of these signals differ.

> Document your observations regarding how the relative audio properties of the signal change as you vary the amplitude, frequency, and waveshape of the signal. Be specific. If you double the frequency, what happens to the pitch? If you decrease the amplitude to half of its original value, is the result half as loud? How does the perceived audio character of the signal at a fixed frequency change with the changing waveshape?

Problem 2.   Playback of a Sampled Signal

The file 'P_2_1.wav' contains a segment of a sampled signal. Play this file using your computer's sound programs, and play the same file using the **sound** command from inside of MATLAB. You will need to use **readwav** to load in the sound file.

> A brief note about **sound**: MATLAB attempts to scale your vector for playback. If you get distorted, unintelligible sounds, you probably need to correct the autoscaling. If you type **saxis('auto')** before each **sound** command, the scaling will reset the minimum and maximum values for playback. You should always scale your vector to be in either an 8-bit (-127 to 128) or 16-bit (-32767 to 32768) range before storing as a sound file.

> What do you hear? Can you identify it? Comment on the quality.

Problem 3.   Sampling and Playback of Signals

In this part of the experiment, you will learn how to digitize a signal and examine the resulting data with MATLAB. Be sure to save the signals that you sample in this section as vectors in MATLAB as you will use them in the next section. NOTE: Using short segments of digitized signals will allow you to proceed much more quickly in the lab.

1. Generate a 100-Hz sine wave with the function generator. A signal amplitude between 0.5 and 0.7 V is strongly recommended (verify with the oscilloscope before connecting input to the card!).

2. Use your computer's sound software to sample the sine wave (a sampling frequency of 8000 Hz will be sufficient).

3. Play back your waveform and view it on the oscilloscope.

4. Use the **readwav** command from inside MATLAB to import your .wav file and make a plot of the signal.

5. Now repeat steps 2-4 except sample your own speech instead of the sine wave.

> Generate plots of segments of your tone and speech signals. Be sure to label your axes. Indicate how features on the plot correspond to sounds in your speech sample.

Problem 4.   Some Audio Effects

Apply each of the following MATLAB functions to both the sine wave and speech segments that you digitized in the previous section. Store your new signal to a .wav file using 'writewav' and then play back the signal. Observe the effect of these digital manipulations on the audio properties of the signal.

1. The MATLAB function **half** created by you in the first laboratory.
2. The MATLAB function **double** created by you in the first laboratory.
3. The MATLAB functions **fliplr** and **flipud.**

> Document how the MATLAB functions you use change the properties of the original tone and speech signals. Compare and contrast the relative impact that these functions have on the audio properties of the speech and tone. Does the speech segment sound like it was produced by the same speaker as the original? Why or why not?

Problem 5.   Echo Effects

Use the MATLAB function file you created to create an echo signal using your original speech signal and a constant attenuation. You will need to create a signal that is a scaled version of the original and add it to the original with the desired offset in time. Experiment with different values for the time delay and attenuation. Listen to the resulting signals and note your observations.

1. Generate a 0.25-second echo effect. Let α=0.65.
2. Use a non-constant attenuation function, such as an exponential: $\alpha(t) = Ae^{-t/\tau}$. Determine values of A and τ that make an interesting effect.
3. Use several different delays and amplitudes to get a variety of echoes.
4. Use an oscillatory attenuation function, such as $\alpha(t)=A\cos(\omega t)$.

> Describe the impact that your choice of time delay T and attenuation α have on the resulting synthesized echo. Be specific about changes in parameters and relative impact on sound. What happens if T changes during the echo-generation process?

Laboratory Assignment 3

Digital Music Synthesis:
Beethoven's Fifth Symphony Using MATLAB

PURPOSE

In this laboratory assignment, you will use MATLAB to synthesize the audio tones that make up a well-known sequence from Beethoven's Fifth Symphony by using basic mathematical functions as fundamental building blocks. You will explore how mathematical operations on these signals yield improvements in perceived quality and can be used to generate special effects. By generating a spectrogram of the signal, you will learn how to represent changes in the frequency content of signals over time and how to interpret such spectrograms.

3.1 OBJECTIVES

By the end of this assignment, you should be able to:

Time Domain

1. Construct physically meaningful signals mathematically in MATLAB using the basic signal building blocks and operations on these signals.

2. Gain physical understanding of constructed signals by viewing them on the oscilloscope and using audio playback.

3. Improve the perceived audio quality, shift pitch and generate special effects using mathematical operations on signals.

Frequency Domain

4. Generate a time/frequency representation of a signal and understand the information contained therein.

5. Explain how the spectrograph and musical score represent the same information.

6. Predict how time-domain changes to the signal that produce special effects will alter the spectrograph.

3.2 REFERENCE

Review Topics

    1. Signal representations, particularly sinusoids and tones

    2. Sampling, interpolation and decimation from Lab 2

    3. Fourier series

Exploratory Topics

    1. Special effects

    2. Fourier analysis

    3. Time / frequency representations

    4. Musical notes and practices

3.3 LABORATORY PREPARATION

Problems

Question 1.   Using the information in the background section, write a mathematical expression for a CT signal $c(t)$ that represents your composition. Use shifted CT unit step functions to express your signal analytically as a single function. Be sure to pay attention to the short pauses after the notes (otherwise, the three G notes will sound like one long tone), the frequencies of the notes, and the durations of the notes and rests. For simplicity, you can assume that the amplitude (volume) for each note is constant over its duration.

Question 2.   Using your answer to Question 1, write a mathematical expression for the DT signal $c[n]$ obtained by sampling the CT signal $c(t)$ at 8 kHz. Be sure to convert your CT unit step functions to valid DT unit step functions.

Question 3.   Determine analytical expressions for the operations performed by your functions **half** and **double** and the MATLAB function **flipud**. Assume that your original signal $x(t)$ is sampled and stored as a column vector **x** and that the operations result in a column vector **y**, which results in the signal $y(t)$ for audio playback. Express $y[n]$ using time scaling and shifting of $x[n]$, and express $y(t)$ using time scaling and shifting of $x(t)$.

Question 4.   Determine how you would use MATLAB to generate the signal in Question 2 Also indicate how you would need to modify this MATLAB function to cause the volume of each note to decay exponentially with time.

Question 5.   Find mathematical expressions for some of the effects outlined in the section on volume variations on page 29 that you intend to use. Use simple examples to illustrate your answers.

Question 6.   How do you expect the volume variations you discussed in Questions 4 and 5 to impact the signal spectrum? What will be the impact on the spectrograph?

Question 7.   Given the signal $x(t)$ consisting of two time-varying tones, sketch the spectrograph

$$x(t) = [\cos(100t) + 2\sin(\omega_1(t) t + \pi/4)] u(t)$$
$$\omega_1(t) = 100(1 - 2e^{-10t})$$

3.4 BACKGROUND

In this section, we explore how to use simple tones to compose a segment of music. By using tones of various frequencies, you will construct the first few bars of Beethoven's famous piece Symphony No. 5 in C-Minor. Each musical note can be represented by a sinusoid whose frequency depends on the note pitch.

Frequencies in Music

Musical notes are arranged in groups of twelve, called *octaves*. The notes that we'll be using are in the octave containing frequencies ranging from 220Hz to 440Hz. The twelve notes in each octave are logarithmically spaced in frequency, with each note frequency being $2^{1/12}$ times the frequency of the next lowest note. Thus, a 1-octave pitch shift upwards corresponds to a doubling of the frequencies of the notes in the original octave. Table 2 shows the ordering of notes in the octave to be used to synthesize the music for this part of the experiment as well as the fundamental frequencies for those notes.

Table 2: Notes in the 220-440Hz octave

Note	Frequency (Hz)
A	220
$A^\#$, B^b	$220*2^{1/12}$
B	$220*2^{2/12}$
C	$220*2^{3/12}$
$C^\#$, D^b	
D	
$D^\#$, E^b	
E	
F	
$F^\#$, G^b	
G	
$G^\#$, A^b	$220*2^{11/12}$

A musical score is essentially a program of sorts—a plot of frequencies (notes, on the vertical scale) versus time (measures, on the horizontal scale). The musical sequence of notes to the piece you will synthesize is shown in Figure 3.4.1.

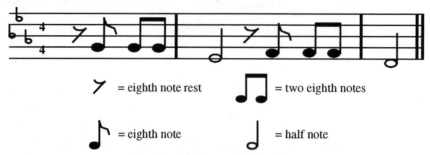

γ = eighth note rest   = two eighth notes

= eighth note   = half note

Figure 3.4.1 *Musical Score for Beethoven's Fifth*

Musical Notation

The horizontal lines on the scale in Figure 3.4.1 represent the notes E, G, B, D, and F from bottom to top. The spaces between the lines are used to represent the notes F, A, C, and E, again from bottom to top. Note that A-G only yields seven notes.

The additional changes in pitch are denoted by adding the symbols $^\#$ (sharp: increase pitch by $2^{1/12}$) or ^{b} (flat: decrease pitch by $2^{1/12}$) to a given note.

In the musical score in Figure 3.4.1, the first three eighth notes are all G. While it appears that the first half note should be an E, it is actually an E^b due to the inclusion of the three flat symbols at the left of the score. The next three eighth notes are all F, and the final half note is D. You can get the fundamental frequencies for these notes by following the pattern given in Table 2.

Note Durations

In the simplest case, each note may be represented by a burst of a sinusoidal tone followed by a shorter period of silence (a pause). The pause allows us to distinguish between separate notes at the same pitch.

The duration of each tone burst is determined by whether the note is a whole note, half note, quarter note, eighth note, etc. Obviously, a quarter note has twice the duration of an eighth note and so on. The short pause following each note should be of the same length regardless of the note's duration. In composed music, longer periods of silence that are part of the musical score are indicated by one or more rest symbols. The two "4" symbols at the left of the score indicate that there are 4 beats per measure in the score and that a quarter note lasts one beat. For this particular piece of music, the duration of a beat should be about 1/2 second.

Time / Frequency Representations of Signals

Creating music from a musical score is a straightforward process. What about creating a musical score from music? There has to be a way to determine the notes in a sampled signal and to find out when they begin and end.

Such a process is called a time / frequency representation of the signal, or a spectrograph. Time is plotted on one axis, frequency on another, and color is used to represent the presence or absence of frequencies in the signal during that particular time. This is easily done in MATLAB by breaking the sampled signal into small time segments, of approximately 50 msec in duration, and plotting the energy present at each frequency for that time segment. This process is performed by the file **P_3_9.m**; check the help on this function for specific details of its operation.

Note that the information displayed by this function is almost identical to a musical score. There are some errors in what is displayed, since evaluating the frequency information for a small piece of data is inaccurate.

Improving Perceived Quality

Volume variations: Typically, when a note is played, the volume rises quickly from zero and then decays over time, depending on how hard the key is struck and how long it is depressed. The variation of the volume over time is divided into four segments: Attack, Decay, Sustain, and Release (ADSR). For a given tone, volume changes can be achieved by multiplying a sinusoid by another function called a windowing function. A decaying exponential is the simplest way to modulate the tone volume. You can try concatenating different functions to model ADSR.

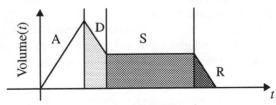

Figure 3.4.2 *An ADSR Envelope*

Overlapping tones: Consider a beginning piano player's attempts at playing a simple tune. This student generally will quit pressing down one key before pressing down the next. A more advanced player would have some notes overlapping (either by using floor pedals or by pressing keys simultaneously). As the volume of one note is decaying, another note is played. Mathematically this can be accomplished by allowing the time regions occupied by different sinusoids to overlap. This will yield a much smoother, less staccato-sounding piece.

Harmonics: While tones are represented by sinusoids, in real instruments the vibrations that generate a given note also include harmonic components. The frequency of the note is called the fundamental frequency f_0; the nth harmonic is a sinusoid having frequency equal to n times the fundamental frequency: nf_0. The amplitude, or power, of the harmonic components is generally less than that of the fundamental and decreases with increasing frequency. For each note, adding in some lower-power harmonics makes the note sound "richer," whereas a single tone sounds very crisp. A word of caution: adding in harmonics of equal or greater power will have undesired effects, such as increasing the perceived note fundamental, and may make the note sound harsh.

To understand harmonics, recall the Fourier series from mathematics; evaluating a series for a certain number of harmonic components is the desired operation. The magnitude of a Fourier series coefficient indicates the energy in a harmonic frequency. Consider the Fourier series for a square wave—it has only odd harmonics and the magnitude of the nth harmonic is proportional to $1/n$.

Reverb: Instruments are usually played in rooms that generate reverberations. Reverberations are reflections off of room surfaces, including walls, ceiling, floors, and obstructions. In DT mathematics, you can view this as generating echoes of the original signal, with decreasing amplitude as the delay time increases. However, a major surface (e.g., a back wall), may generate more power (amplitude) at specific delays. You can determine the delay time by dividing the distance sound travels by the speed of sound (~300 m/s at sea level). A model for generating an echo is given in Laboratory Assignment 2.

3.5 LABORATORY ASSIGNMENT

Part I: Song Creation

Problem 1.   Using MATLAB, construct a row vector that is a discrete-time representation of the notes and rests that make up the musical piece in Figure 3.4.1. Let the sampling rate be 8 kHz for this particular composition. Be sure to pay attention to the short pauses after the notes (otherwise, the three G notes will sound like one long tone), the frequencies of the notes, and the durations of the notes and rests. Once you have a piece that you think accurately represents the music, play the composition. Remember to scale your signal before saving it as a sound file.

> Verify that the timing and pitch of your composition is an accurate representation of the musical score. Describe how it sounds and how it can be improved. Think about terms like richness of sound, clarity of pitch, flow of composition, musical expression, instrumental versus vocal sounds, etc.

Problem 2.   Using what you know about digital pitch shifting and the logarithmic frequency relation between the notes in Table 2, modify your composition using **half** and **double** so that it is played back one octave higher and one octave lower. Note that in the process of digital pitch shifting, you also changed the duration of the signals.

> Discuss what happens to the pitch and duration of a signal when you use your functions **half** and **double** to modify it. Be specific: indicate new durations in seconds and new pitches by both frequency and musical pitch.

Problem 3.   Try to have the volume of each note decay over time to make your music sound more interesting and realistic.

> How did changing the signal amplitude affect your perception of the volume? If you double the signal amplitude, does the volume double as well?

Part II: Note Modifications

Problem 4.   Modify your pitch-shifted composition so that each note is of the same duration as in the original composition.

Problem 5.   Can you pitch shift your composition up by a single half-step? Consider interpolating first and then resampling.

Problem 6.   Allow decaying notes to overlap slightly in time.

Problem 7.   Try adding harmonics to create a richer sound.

Problem 8.   Use the M-file you created in Lab 1 to generate an echo to create reverb effects.

> For each of the possible improvements above, determine the mathematical expressions for the sequence of operations needed to achieve the desired result. Use a simple mathematical example to illustrate your reasoning. Are you able to achieve the desired result? Discuss what worked well and what needs improvement. Explain why.

Part III: Frequency Analysis

Problem 9.   Compute the time-frequency representation of your signal from Problem 1 with the function **P_3_9.m**. Compare the plot with the musical score in Figure 3.4.1, and identify each note from the score on the spectrogram.

> The spectrogram has time in seconds on the horizontal axis and frequency in Hz on the vertical. Musical note durations should be clearly evident, as should the frequencies at which they occur. Does your synthesized note sequence accurately represent the musical score?

Problem 10.   Generate spectrograms of your pitch-shifted composition from Problem 2. You will need to change the frequency range displayed by **P_3_9.m** to see any results.

> How do these differ from the spectrogram of the original composition? Verify that your results are as expected by identifying the pitch and duration of each note on the spectrogram.

Problem 11.   Generate spectrograms for some of the special effects you generated in Problems 4—8.

How do these spectral effects and improvements alter the spectrogram? Based on the operations you performed and properties of the Fourier Transform, do these changes make sense?

Laboratory Assignment 4

Fourier Sound Synthesis

PURPOSE

This lab investigates how to use a computer to evaluate the Fourier series for periodic signals and to synthesize audio signals from Fourier series coefficients. The Fourier series should be a familiar mathematical concept. We show how to use the DFT (Discrete Fourier transform) to compute the Fourier series for digital sound synthesis. Using a computer, we can determine the Fourier series coefficients for an arbitrary periodic signal corresponding to a sampled audio signal having a given pitch. Then similarly shaped waveforms having different fundamental frequencies, i.e., different pitch, can be generated to create interesting musical sounds. We explore how these sounds can be modified using ideal filters.

4.1 OBJECTIVES

By the end of this laboratory assignment, you should be able to:

1. Synthesize a signal from a finite number of Fourier series coefficients.

2. Compute Fourier series coefficients for an arbitrary periodic signal (that may not have an analytical solution).

3. Perform and analyze basic ideal filtering operations on periodic signals.

4.2 REFERENCES

Review Topics

    1. Fourier series analysis and synthesis

Exploratory Topics

    1. Discrete-time Fourier analysis and synthesis

    2. Ideal filtering of arbitrary periodic waveforms

Reference

    1. Deller, J. R. Jr. *Tom, Dick, and Mary Discover the DFT.* IEEE Signal Processing Magazine, April 1994 pg. 36-50

4.3 LABORATORY PREPARATION

Problems

Question 1.   Assume that a significant harmonic is defined to have a magnitude within 50 dB of the magnitude of the highest harmonic. Using this figure of merit, how many harmonics are significant for a square wave? Use the analytic formulation for the FS of a square wave. You may wish to review the definition of dB given in Laboratory Assignment 3.

Question 2.   Consider a signal that comprises only two harmonic components: the third harmonic, having a magnitude of two and an angle of $\pi/4$, and the fifth harmonic, having a magnitude of 1/2 and an angle of $3\pi/4$. Construct 2 periods of this waveform, assuming that its fundamental frequency is 50 Hz. If an ideal lowpass filter is created to remove the fifth harmonic, what would be a reasonable cutoff frequency (in Hz)?

Question 3.   Given the waveform $f(t)$ below, determine its Fourier series coefficients a_n analytically in terms of the integer index n, assuming it is periodic outside of the interval shown.

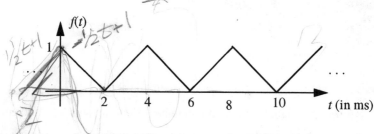

Figure 4.3.1 *Waveform for Problem 3*

What are the Fourier series coefficients for the signal resulting from filtering $f(t)$ with an ideal LPF that has a cutoff frequency of $3.5\omega_0$, where ω_0 corresponds to the fundamental frequency of $f(t)$?

Question 4.   Verify that your Fourier series analysis from Question 3 is correct by plotting the waveform resulting from using the first four Fourier series coefficients. What happens to the waveform shape as you increase and decrease the number of Fourier series coefficients used in the approximation? Identify any differences between the Fourier series approximation and the original signal. Hypothesize, on the basis of your observations, as to whether these differences will disappear if enough of the harmonic frequencies are included in the approximation. Plot the magnitude of the first four Fourier series coefficients as a discrete-time sequence with the corresponding harmonic frequencies indicated on the horizontal axis.

Question 5.   Discuss whether the Fourier synthesis method described for the guitar could be applied to different types of instruments, e.g., a trumpet. Explain your reasoning.

■

Motivation for Frequency Analysis

Much of signal processing, communication and control systems analysis relies upon frequency analysis of relevant signals and systems. Frequency response, bandwidth, and tone control on stereo systems are all concepts based on the idea that a signal can be thought of as comprising sinusoidal components at different frequencies.

Traditional frequency analysis uses two basic transforms: the Fourier series, for periodic waveforms, and the Fourier transform, for aperiodic waveforms. In both cases the information from a signal is converted to a frequency-domain representation; instead of representing the information in the signal as a function of time, it is represented as a function of frequency.

What does plotting the signal content as a function of frequency tell us? In a broad sense, it tells us at what frequencies the signal has energy, just as a prism visually shows the colors—frequencies—present in a light source, or a stereo spectrum analyzer shows the relative loudness of tones in a musical piece. Frequency analysis helps to understand other technologies as well. For example, the electric power in a standard wall outlet contains energy predominately at 60 Hz— AC power. A radio station broadcasts energy around its transmission frequency, which is indicated on your radio station dial. By transforming signal data to be viewed in the frequency domain, we can find out what frequencies are significant in a given signal.

The behavior of systems also can be described using frequency-domain concepts. Systems are characterized by their frequency response, i.e., how they affect the frequency content of an input signal. For example, the tone control on a stereo (bass and treble knobs) has two effects: the bass knob controls the gain (relative loudness) on low-frequency signal components, whereas the treble knob controls the gain on high-frequency signal components. Examining the frequency content of a stereo signal before and after tone control can tell what the two knobs are doing. In a more general sense, the frequency response of any unknown system can be determined from the relative frequency content of signals at its input and output.

This set of experiments deals with three concepts in the context of synthesizing and modifying audio sounds: calculation of the frequency information in arbitrary periodic signals via the Fourier series (FS); the effect of certain systems, called ideal filters, on the FS coefficients and the audio properties of signals; and determination of a system's frequency response based on input and output signal Fourier series spectra.

Why bother with the Fourier series at all? This mathematical representation provides a means for determining the frequency content of periodic signals. For a given periodic signal, the gain (magnitude) and time-shift (phase) of sinusoids at each harmonic frequency define the signal's frequency spectrum. Knowing the frequency content of a signal allows you to determine how a signal will or should be modified by a physical system. In this lab, we investigate how systems affect the frequency content of a signal in the context of ideal filters having both periodic input and output signals that can be represented by their respective Fourier series.

Scenario

You are interested in getting hired by a small start-up company that develops multimedia application packages for entertainment and education. This company uses advanced digital audio
processing in their products to produce high quality sound with low storage requirements,
which is the main selling point for their products. You have researched the company's products
and financial status and are impressed. This company is interviewing on campus. Knowing that
the competition is stiff for the single engineering position available, you would like to make a
good impression.

You have played around some with sound synthesis using simple tones and sampled sounds, but
recognize the large storage requirements needed to produce high-quality music using only samples of digitized instrument sounds. You have heard of something called Fourier synthesis.
Recalling the Fourier series from your coursework, you wonder if these mathematical techniques are used in Fourier synthesis to develop methods for low-cost, high-quality digital sound
synthesis. So you decide to explore how to use a computer to perform Fourier series analysis
and synthesis to see what kinds of sounds you can create.

Fourier Series and Discrete Fourier Series

The Fourier series representation of a periodic signal should be a familiar concept from introductory engineering coursework. The main concept behind the FS is that periodic signals can
be represented as a sum of sinusoidal signals, with the restriction that the resulting sum has the
same fundamental frequency as the signal itself— i.e., the same periodicity. A classic example
is the construction of a square wave from the FS coefficients; there is a MATLAB demo that
does just this: type **fourier** in MATLAB to see the development.

The FS is defined by two equations: an analysis equation, which, from the signal, determines
the FS coefficients, and a synthesis equation, which defines how to construct the signal as a sum
of sinusoids weighted by the FS coefficients.

1. Fourier Series: Analysis
$$a_k = \frac{1}{T} \int_{-T/2}^{T/2} x(t) e^{-j2\pi k f_0 t} dt$$

2. Fourier Series: Synthesis
$$x(t) = \sum_{k=-\infty}^{\infty} a_k e^{j2\pi k f_0 t}$$

In both equations (1) and (2), $f_0 = 1/T$, the fundamental frequency of the signal, T is the period
of the signal, and the a_k's are the complex coefficients, each of which weights the kth harmonic
(having frequency $k f_0$). Note that since the above synthesis equation is expressed in terms of
complex exponential functions, the index on the summation includes negative values so that the
resulting sum defines sinusoids. (Recall that $\cos(\theta) = (e^{j\theta} + e^{-j\theta})/2$.) Often the sign associated
with the index is associated with the frequency; $k f_0$, where k is negative results in a mathematical artifact called a negative frequency.

Since the signals in which we are interested are real-valued functions of time, the Fourier coefficients form complex conjugate pairs, $a_k = (a_{-k})^*$.

The Fourier series also can be expressed as a sum of sinusoids,

$$x(t) = a_0 + 2 \sum_{k=1}^{\infty} |a_k| \cos(2\pi k f_0 t + \angle a_k)$$

Note that the magnitude of each Fourier series coefficient determines the amplitude of the sinusoid at the corresponding harmonic frequency, and the phase of the FS determines the phase shift of the sinusoid.

If the signal is sampled to yield a discrete-time signal, we can use the discrete Fourier Series for analysis and synthesis. In this case, integration is replaced by summation in the FS analysis equation, and the signal periodicity is expressed in terms of an integer number of samples N rather than time T.

3. Discrete Fourier Series: Analysis

$$a_k = \sum_{n=0}^{N-1} x[n] e^{-j2\pi kn/N}$$

4. Discrete Fourier Series: Synthesis

$$x[n] = \frac{1}{N} \sum_{k=0}^{N-1} a_k e^{j2\pi kn/N}$$

Note that in the discrete case only a finite number of DFS coefficients are used. Since the data is sampled, there is a restriction on the frequencies that can be represented, limited to half of the sampling rate. This gives a maximum discrete-time frequency of $\Omega = \pi k N$, corresponding to a continuous-time frequency of $\omega = \Omega/T_s$ where T_s is the sampling period. The remaining terms in the sum correspond to the negative frequencies values, shifted by $+2\pi$, as needed to create sinusoids from complex exponentials.

These two equations are readily implemented in MATLAB: simply write a script that evaluates the sums for all n and k values. There are, however, more efficient methods for computing the DFS using Fast Fourier Transforms (FFT), which will be discussed later.

Interpreting Fourier Series

So now we have a set of Fourier series coefficients for a signal, and we can recreate the signal from its coefficients. What does this mean?

We are interested in knowing a few things about a signal: what range of frequencies does it contain? Which of these have the most energy? How much energy is stored in a particular frequency? How can the signal be changed by altering its frequency content? The FS representation provides us with a means for answering these questions.

To help answer these questions, we can plot the magnitude of the FS coefficients as a DT sequence where the horizontal axis is used to indicate the harmonic frequency ($k f_0$) associated with each coefficient. Consider the FS representation of a periodic square wave, which is commonly used as an example in lecture or as homework. It contains only odd harmonics ($k = 1,3,5$, etc.) and the amplitude of each harmonic is proportional to the frequency ($1/k$). The frequency spectrum is plotted in Figure 4.4.1 for a 5-Hz periodic square wave.

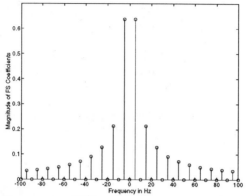

Figure 4.4.1 *FS Coefficients for Square Wave*

It is clear from Figure 4.4.1 that most of the energy is contained in the fundamental or first harmonic, which is at 5 Hz. The energy at each higher harmonic drops off rapidly. If we were to truncate this series by using only the first 10 harmonics to represent the signal, all of the energy above 55 Hz would be lost.

Digital Fourier Series Analysis and Synthesis

Since we will be using MATLAB on a computer, we will focus on how to compute Fourier spectra for discrete-time signals resulting from sampling continuous-time signals. This process is facilitated by the DFT (Discrete Fourier Transform), implemented in MATLAB using the command **fft**, which uses the computationally efficient method called the Fast Fourier Transform (FFT).

The Fourier transform is an analytical method for determining the frequency content of continuous-time analytic signals, which may or may not be periodic. Since aperiodic signals do not have a fundamental period, they may have power at a continuum of frequencies, rather than only at harmonic frequencies. Thus all possible frequencies can be considered in analysis and synthesis. The continuous-time FT is an integral evaluated over the entire signal, from the beginning to the end of time.

However, to use a computer to evaluate the spectrum (Fourier transform) of a signal represented by a vector of time-domain samples, it is neither possible nor desirable to numerically evaluate this integral for all time. Instead of approximating the Fourier transform integral by a summation over infinite time, we only sum over a finite number of samples. However, the number of samples used in the time domain corresponds to the number of "samples" of the signal's continuous frequency spectrum. The frequency resolution of the FT is limited to the number of points used in the computation. If we want to analyze for higher frequencies, we must sample at a higher rate since the uppermost frequency is limited by the sampling rate; if we want to analyze more densely, i.e., have smaller frequency steps, we must take a longer time sequence.

The analysis and synthesis equations that result from this sampling of the signal and its spectrum define what is called the Discrete Fourier Transform:

5. DFT: Analysis

$$X[k] = \sum_{n=0}^{N-1} x[n] e^{-j2\pi nk/N}$$

6. DFT: Synthesis

$$x[n] = \frac{1}{N} \sum_{k=0}^{N-1} X[k] e^{j2\pi nk/N}$$

In these equations, $2\pi nk/N \equiv \omega_k$. Note that these equations look virtually identical to those defining the DFS. The only major difference is that the frequencies corresponding to each sampled Fourier transform value, or Fourier coefficient $X[k]$, are not necessarily related to a fundamental period. To calculate the DFS coefficients using the DFT, we need to define N such that it corresponds as closely as possible to one period of the signal, which is defined by the sample values stored in $x[n]$.

The DFT in MATLAB is not computed using loops that implement the summation equations above; rather, the Fast Fourier Transform (FFT) is used. The FFT essentially streamlines the DFT computations to have a minimal number of multiplications by removing redundancy from the DFT calculation. The MATLAB implementation of the FFT automatically chooses the most efficient way to perform the calculation based on the length N selected for the transform.

Fourier Analysis Using MATLAB

Assume you have a signal $x[n]$ in the vector **x**. The DFT of the signal is computed and stored in **X** by entering **X = fft(x)**. The first value, **X(1)**, has the value of a_0, corresponding to the "average" of the signal.

The best way to understand how to interpret the results you obtain using MATLAB is via example. You can run the examples by simply typing the filename indicated. We suggest that you explore and understand these examples prior to attempting the laboratory assignment.

E_4_1.m: In this example, the DFT of a sum of two cosines is displayed as a function of frequency. Centering about an integer index of zero is accomplished using **fftshift**.

E_4_2.m: In this example, the DFT of a length-N vector **x** is computed using different length DFTs. The vector **x** represents the DT signal $x[n]$ obtained by sampling the signal $x(t) = \cos(20\pi t)$ every nT_s seconds, where $T_s = 0.01$ s. The DFTs computed using different lengths N, corresponding respectively to (1) five periods of data, (2) five periods of data padded with zeros, and (3) 4.5 periods of data, are computed. You should examine the results and note the impact of the DFT length. When you are familiar with Fourier transform properties, you may wish to revisit this example and see if you can analytically understand these differences.

Fourier Series Analysis of Real Signals

The Fourier series is a good analytical tool, and students always get to solve interesting problems like finding the Fourier series of a square wave or a triangle wave. What about using the Fourier series on a real signal?

Let's look at analyzing the vibrations of an acoustic guitar string. If a string is plucked at a single point, we expect to see a triangular shape appear on the string, with the highest point occurring where the string was plucked. By doing some analysis of this ideal vibrating string, the Fourier Series coefficients are

$$C_n = \frac{L^2 h}{2n^2 \pi^2} \sin\left(\frac{n\pi}{L}\right)$$

where L is the proportion of the string length from the point where the string was plucked to the bridge (e.g., 1/5) and h is the initial displacement. It should be noted that the C_n values fall off at $1/n^2$, with zeros at the Lth harmonic. Since the pluck point occurs at $1/L$, no energy is imparted to that particular frequency (and its harmonics). In Figure 4.4.2 the string shape for a pluck point of 1/4 is shown (with a period of 225 samples).

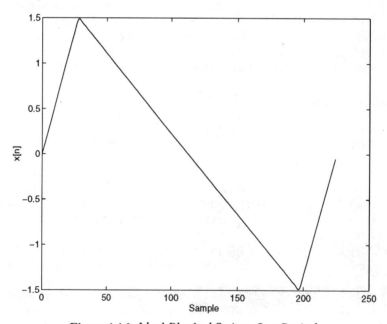

Figure 4.4.2 *Ideal Plucked String, One Period*

If we sample the sound resulting from plucking an acoustic guitar string, however, we will not see a triangular waveform. Instead, we will see a waveform that looks like it could be a triangle wave, but is smoothed out and has some distortions and bumps, as shown in Figure 4.4.3.

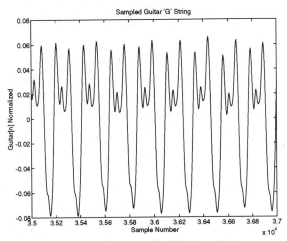

Figure 4.4.3 *A Sampled Guitar String*

This signal is still periodic, so we can find a Fourier series representation for it and compare it to the Fourier series for an ideal plucked string to see what impact the guitar body has on the waveform. In order to do this, we have to extract a single period of the signal. Once we have that, we can apply **fft** to find the FS coefficients.

We can then pose the question "How does the body change the sound of a plucked string from that of an ideal?" Equivalently we could ask, "How does the guitar body modify the harmonics of the plucked string?" The impact of a system on the frequency content of a signal is called its frequency response, and the process of modifying this frequency content is called filtering.

Filtering Operations

To determine the frequency response of the guitar, we are interested in how the magnitudes of the input FS coefficients are modified by multiplicative gains, and how the phases of the FS coefficients are modified by additive terms. If we assume that the effect of the guitar body can be roughly approximated by a CT linear, time-invariant system, then the FS for the periodic output signal $y(t)$ can only contain the same harmonic frequencies as the input $x(t)$; thus it can be expressed as the Fourier series given below, where the a_k are the FS coefficients and T_0 is the fundamental period of $x(t)$

$$y(t) \; = \; H(j0)\, a_0 + 2 \sum_{n=1}^{\infty} |H(j\omega_n)||a_n| \cos\left(\frac{2\pi n}{T_0} t + \angle a_n + \angle H(j\omega_n)\right)$$

The function $H(j\omega)$ is called the system frequency response. Note that the magnitude and phase of the frequency response, when evaluated at the harmonic frequencies, determine the gain and phase shift, respectively, that are introduced by the system at each harmonic. Thus only the magnitude and phase of the input FS terms will be modified by the filter, and the amount of the modification depends upon the frequency. Note that if the equation above is viewed as the Fourier series for the output signal $y(t)$, then the coefficients for $y(t)$ are related to those for $x(t)$ by the complex multiplication $c_{ny} = H(j\omega_n) c_n$.

Filters generally are defined by which frequencies they allow to pass from input to output, or equivalently by those that are attenuated. For example, lowpass filters pass low frequencies and attenuate high frequencies. Shown below are example plots of the magnitude $|H(j\omega)|$ vs. ω for the different types of ideal filters. The filter gain is defined to be G, ω_c is the cutoff frequency for high- and lowpass filters, and ω_h and ω_l are the cutoff frequencies for bandpass filters. PB stands for passband, defined to be the range of frequencies which are passed through to the output, and SB stands for stopband, the range of frequencies removed.

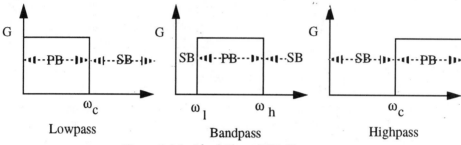

Figure 4.4.4 *Ideal filters $|H(j\omega)|$ vs. ω*

For the ideal filters shown, the magnitude of the filter frequency response, $|H(j\omega)|$ is equal to the gain G in the passband and zero in the stopband. The phase of the frequency response is assumed to be zero, which implies that the filter does not cause any time shift.

4.5 LABORATORY EXPERIMENT

Evaluating the DFS and IDFS

Problem 1.   Write a MATLAB function to evaluate the DFS of a signal using **fft**. Test your function by evaluating the FS coefficients of both a 50-Hz square wave and a 50-Hz sine wave. Use an 8 kHz sampling rate and a value of N corresponding to exactly 1 period.

> Plot the FS coefficients vs. frequency. Verify that your results are correct by comparing your results to the expected analytical results.

Problem 2.   Write a MATLAB function that can synthesize a signal from FS coefficients given the fundamental frequency. Attempt to generate the signals from Problem 1 using your function. You will probably need to use **real** if you use **ifft** to remove small imaginary components generated by roundoff.

> What happens as you increase the number of FS coefficients used to synthesize the signal? Does this observation make sense given your theoretical understanding of FS?

Problem 3. We know that the frequencies present in a signal, and therefore in its FS, are harmonically related to the fundamental frequency f_0. What happens if, after analyzing a given periodic waveform, we change f_0 when we resynthesize the waveform to essentially replicate the waveform "shape" at a different pitch? Determine the first 12 FS coefficients for a 50-Hz sawtooth wave, then resynthesize using $f_0=100$ Hz.

> What similarities and differences do you observe between the original and resynthesized signals? Consider both time and frequency domains.

Problem 4. In the file **P_4_4.mat** are two periodic waveforms, **x** and **y**; **x** has been filtered with an ideal lowpass filter to generate **y**. By looking at the FS coefficients, determine the cutoff frequency, in radians per second, and gain of the lowpass filter.

> If you assume a sampling rate of 44.1 kHz, what harmonic frequencies are present? What is the filter cutoff frequency as a function of f_0? How exact is your estimate of the cutoff frequency (i.e., in what range of values could it be)?

Problem 5. Construct a MATLAB function that plots the magnitude of the Fourier series coefficients for a periodic square wave as a function of frequency. Normalize by dividing the magnitude of all coefficients by a_0. Fix the width of the rectangular pulses to be $A = 1$ s, and let the period T be a variable. Assume a sampling rate of 100Hz. Generate the plot for $T = 2$ s. Repeat the process several times, doubling the period each time.

> Plot your results using the same frequency scale. What do you observe? What might be the implications of these results on the accuracy of the method you used in Problem 3 to resynthesize using a different fundamental frequency?

Some Aspects of DFS

Problem 6. Analyzing Acoustic Guitar Data

In **P_4_6.mat**, a waveform corresponding to an ideal plucked string is stored in vector **x_ideal**. The signal resulting from sampling the sound of an acoustic guitar in response to a plucked string is stored in **x_sampled**. Compute the Fourier series coefficients for **x_ideal**, and contrast the results to the mathematical formula on page 40. Determine the pluck point L. Compute the Fourier series coefficients for **x_sampled**, and compare the Fourier series coefficients of **x_ideal** and **x_sampled**. Calculate the change in magnitude and phase for each coefficient from ideal to sampled data, and plot these values. What do these tell you about the frequency response of the guitar body? Create a function that synthesizes the sampled guitar sound from the FS coefficients for the ideal waveform, given the magnitude and phase changes found above.

> If you had to design a filter that implemented the changes, how would you do it? For which pitches (i.e., at what frequencies) can you be sure that your filter accurately models the guitar's response?

Problem 7.   Creating Beethoven's Fifth Using Fourier Synthesis

In Chapter 3, you synthesized Beethoven's Fifth Symphony by concatenating sine waves of different frequencies. Recreate the first four notes of this musical score by pitch-shifting the sampled guitar string using the method described in Problem 3 and the Fourier series coefficients that you computed in Problem 6.

> Do the notes you've created sound better than the tones you used before? Does it sound like it was played on a guitar? Why or why not?

Problem 8.   Identifying Filter Frequency Responses

In **P_4_8.mat**, there are two periodic waveforms, **x** and **y**; **x** has been filtered, creating **y**. From the FS coefficients of **x** and **y**, determine as much as possible about the frequency response of the filter. Try changing the FS components of **x** with the frequency response you've found, and compare the results to **y**.

> How much do the FS coefficients tell you—do you know the entire frequency response of the filter? How accurate is your determination of the response?

Laboratory Assignment 5

Simulating the Operation of a Prosthetic Limb

PURPOSE

In this laboratory assignment, you will simulate the operation of a prosthetic limb by using a differential equation model. Using MATLAB, you can easily explore how the time-domain response changes as you change different parameters in the human-machine system, and how this response relates to the characteristic root locations and physical operation of the prosthesis. Armed with this information, you are to choose values for tunable system gains to meet the desired design criteria for limb performance. The transform domain section of the laboratory explores how the Laplace transform is used to define the transfer function of the system and how this transfer function concisely represents critical information about the system.

5.1 OBJECTIVES

By the end of this assignment, you should be able to:

Time Domain

1. Predict the time-domain response of CT LTI systems from the characteristic root locations.

2. Use MATLAB to simulate the time-domain response of CT LTI systems modeled by state-space and nth-order differential equations.

3. Identify resonant frequencies and predict system behavior of CT LTI systems.

4. Plot characteristic root values in the complex plane as the feedback gain is increased from zero.

Transform Domain

5. Determine the transfer function of an interconnected system, either from a block diagram or a state-space representation.

6. Generate and interpret root locus plots using MATLAB.

7. Determine and plot system frequency responses and provide physical interpretation of your results.

5.2 REFERENCES

Review Topics

1. State-space representation

2. Input/Output differential equation representation

3. Characteristic roots and responses

4. Definition of zero-input and zero-state response

5. Definition of transfer function

Exploratory Topics

1. Root locus generation and interpretation

2. Transfer function, poles and zeros

3. System stability and poles

4. Resonant frequencies, performance and BIBO stability

5. MATLAB for simulation of I/O or state-space differential equation models

6. Prosthesis modeling and operation

Application Reference

1. Artificial limb model, *Design of Feedback Control Systems* (Hostetter, Savant, Stefani, 1982, section 3.11)

5.3 LABORATORY PREPARATION

Use the MATLAB help facility to look up **lsim**, **ss2tf**, **roots**, **eig**, **rlocus**, and related functions. Note that most of these functions are in the Control Systems toolbox.

Problems

Question 1.   Determine the state-variable representation for the overall system shown in Figure 5.4.1 using the equations provided. Let $d(t)$ be the input signal and $y(t)$ be the output signal. For your state variables, use position $x_1(t) = y(t)$, velocity $x_2(t) = v(t)$, acceleration $x_3(t) = dv/dt$, measurement $x_4(t) = m(t)$, and integrated error $x_5(t) = i(t)$. Explicitly identify A, B, C, and D. Express your answer in terms of K_B, K_T, and T. You will use these matrices with the MATLAB function **lsim** to simulate the system response in your experiment.

Question 2.   Repeat Question 1 for the feed-forward system using the same states, but using $e(t)$ as the input and $y(t)$ as the output. You will use the resulting A, B, C, and D matrices as inputs to the MATLAB function **rlocus**, which automatically plots how the characteristic roots change with K_B.

Question 3.   For the motor/tachometer subsystem having input $m(t)$ and output $v(t)$, write the input/output differential equation. Express the characteristic root locations for this subsystem in terms of K_T. For what value of K_T is this subsystem critically damped?

Question 4.   Identify mathematical expressions that could be used for the input $d(t)$ and initial conditions on the states to represent stepping up, running/walking at a given pace, and taking a foot off a stool and letting it drop. In other words, for each of these activities, determine how you can mathematically specify the physical initial conditions and desired prosthetic limb height.

Question 5.   Look up the MATLAB function **lsim** using the **help** facility. Indicate how you need to pass the information about the state-space model (A, B, C, D), and initial conditions and inputs to this function. Use your state-space model from Question 1, and the activities described in Question 4.

Question 6.   Determine the transfer function of each block in Figure 5.4.1.

Question 7.   What is the closed-loop transfer function for Figure 5.4.1?

Question 8.   What is the open-loop transfer function for Figure 5.4.1?

Question 9.   For K_T at the critically damped value found in Question 2 and $K_B = 28$, determine the system poles and zeros and system frequency response using MATLAB. Identify the resonant frequencies and explain how these can be determined from the system poles and zeroes.

5.4 BACKGROUND

You have been hired by a biomedical engineering firm that specializes in the design and development of high quality artificial limbs (prostheses). The newest product about to be tested on volunteer patients is an artificial leg for people who are missing their lower leg, including the knee joint. The system uses signals from the thigh muscles to control the motion of the lower leg. A block diagram modeling how this prosthesis works with the human brain and central nervous system (CNS) is shown in Figure 5.4.1. Differential equations, shown on the next page, are used to model how this human-machine team works together.

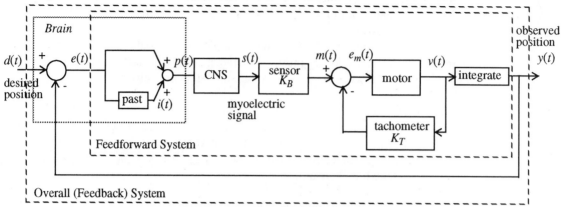

Figure 5.4.1 *Prosthetic Limb System*

Brain

$$e(t) = d(t) - y(t)$$
$$p(t) = e(t) + i(t)$$

$$\frac{di}{dt} = 0.1e\,(t)$$

CNS

$$\frac{ds}{dt} + \frac{1}{T}s\,(t) = \frac{1}{T}p\,(t)$$

Sensor

$$m(t) = K_B\,s(t)$$

Motor/Tachometer System

$$\frac{d^2v}{dt^2} + 11\frac{dv}{dt} + 10v\,(t) = 5e_m\,(t)$$

$$e_m(t) = m(t) - K_T\,v(t)$$

Observed Position

$$\frac{dy}{dt} = v\,(t)$$

Referring to Figure 5.4.1 and the associated equations, the operation of the prosthesis can be described as follows. The input signal $d(t)$ is generated by the brain to be the desired position of the leg in one dimension—i.e., $d(t)$ represents the desired height of the foot at each point in time, and depends on the activity the patient is performing. The output signal $y(t)$ is the observed position (or height) of the foot. The brain uses the present error $e(t)$ between the desired and observed positions and knowledge of past errors $i(t)$ to generate a potential $p(t)$ which is the message signal telling the CNS what action to take. The CNS generates the signal required to carry out the desired action, and this signal is sent to the muscle with some time delay T. The signal at the muscle $s(t)$ (a myoelectric signal) is picked up and amplified by sensors to generate an electrical signal, $m(t)$, representing the velocity required to generate the

Laboratory Assignment 5

desired motion. A tachometer measures the actual velocity generated by the motor $v(t)$ and compares it to the desired velocity $m(t)$. The error between these signals, $e_m(t)$, is used to drive the motor. The new position $y(t)$ is determined from the velocity $v(t)$.

The overall prosthetic limb system here is one of a class of feedback systems. It feeds back the measured, or actual, position for comparison with the desired position. The difference between the observed and desired positions is used to determine an appropriate course of action. In general, feedback has a stabilization effect on systems.

I/O Differential Equations

In general, continuous-time linear, time-invariant systems can be described by nth-order input/output differential equations of the form

$$\frac{d^n y}{dt^n} + a_{n-1} \frac{d^{n-1} y}{dt^{n-1}} + \cdots + a_1 \frac{dy}{dt} + a_0 y(t) = b_0 f(t) + b_1 \frac{df}{dt} + \cdots + b_m \frac{d^m f}{dt^m}$$

where $f(t)$ is the input signal and $y(t)$ is the output signal. This system can be completely characterized by knowing the coefficients a_i and b_j; these coefficients can be passed to MATLAB in the $DEN = [1 \; a_{n-1} \; \cdots \; a_1 \; a_0]$ and $NUM = [b_m \; b_{m-1} \; \cdots \; b_1 \; b_0]$ vectors, respectively, to simulate the system behavior using MATLAB. Note that the ordering is important, as the first element of DEN or NUM is the gain for the highest order derivative, and the last element is the gain for the lowest; be sure to read the MATLAB help for **lsim** carefully. Initial conditions are specified in terms of the values of $y(t)$ and its $(n-1)$st derivatives at the initial time $t = t_0$.

State Space Representations

A system can also be represented by the so-called state-variable description; instead of using one nth-order differential equation, we represent the system by n first-order, cross-coupled differential equations.

$$\underline{x}'(t) = A\underline{x}(t) + Bf(t)$$
$$y(t) = C\underline{x}(t) + Df(t)$$

where the vector $\underline{x}(t) = [x_1(t), x_2(t), ..., x_n(t)]^T$ is called the state vector. Each $x_i(t)$ is a state; $\underline{x}'$ indicates a vector of derivatives of the states; the input and output are as defined in the I/O differential equation; and A, B, C, D are $n \times n$, $n \times 1$, $1 \times n$, and 1×1 (a scalar) matrices, respectively. The set of matrices $\{A, B, C, D\}$ completely specifies the system for MATLAB simulation. In many cases, such as the prosthetic limb problem, this state-space representation is a natural form.

To use the state-space representation in MATLAB, rearrange the given equations so that they can be expressed in the standard form given above. For example, for the motor/tachometer subsystem alone, a state-variable representation could be obtained by defining $m(t)$ to be the input, $v(t)$ the output, and the states as $x_1(t) = v(t)$ and $x_2(t) = dv/dt$. With these definitions, the state-variable representation in standard form is:

$$\begin{bmatrix} \dfrac{dx_1}{dt} \\[2ex] \dfrac{dx_2}{dt} \end{bmatrix} = \begin{bmatrix} 0 & 1 \\ -(10 + 5K_T) & -11 \end{bmatrix} \begin{bmatrix} x_1(t) \\ x_2(t) \end{bmatrix} + \begin{bmatrix} 0 \\ 5 \end{bmatrix} m(t)$$

$$y(t) = \begin{bmatrix} 1 & 0 \end{bmatrix} \begin{bmatrix} x_1(t) \\ x_2(t) \end{bmatrix} + [0] m(t)$$

When using this representation, the n initial conditions are specified in terms of the values of each state $x_i(t)$ at the initial time $t = t_0$.

Transfer Function Representation

The input/output representation for a system can be found from a given state-space representation as follows. Assume that your input and system response are both of the form e^{st}, where s is an arbitrary complex number. Then $d(e^{st})/dt = se^{st}$, $d^2(e^{st})/dt^2 = s^2 e^{st}$, and $d^n(e^{st})/dt^n = s^n e^{st}$. Thus we can represent differential equations using polynomials in s, where we use powers of s as place-markers for derivatives, i.e. $d^n/dt^n \Rightarrow s^n$. Using this approach, we can "solve" the state matrix equations for $x(t)$ in terms of $f(t)$, and then substitute the result into the output equation to "solve" for the output $y(t)$ in terms of s and the input $f(t)$:

$$\frac{y(t)}{f(t)} = C(sI - A)^{-1}B + D = \frac{b_0 + b_1 s + \cdots + b_m s^m}{s^n + a_{n-1}s^{n-1} + \cdots + a_1 s + a_0} = H(s)$$

where I is an $n \times n$ identity matrix. Note that this matrix solution results in a ratio of polynomials with coefficients corresponding to the input/output differential equation. Cross multiplying results in the denominator polynomial multiplied by $y(t)$ and the numerator polynomial multiplied by $f(t)$; substituting derivative operators back for the placeholders, $s^n \Rightarrow d^n/dt^n$ results in the I/O differential equation.

Note that this ratio of polynomials in s is called the transfer function and is denoted by $H(s)$; this is an important signals and systems representation and you will be seeing more of it. The roots of the numerator polynomial are called zeros, and the roots of the denominator, the characteristic roots, are also called poles. In MATLAB, the state-space representation can be converted to a transfer function using the function **ss2tf**, where the notation indicates "state-space to transfer function."

System Characteristic Roots and Responses

What is the response of a system that is modeled by differential equations? Will it oscillate? Will it damp out? In terms of the prosthesis: will it exhibit tremors? Will these tremors disappear over time? The answers to these questions can be found in the characteristic roots of the human-machine prosthetic system model. The characteristic roots are defined to be the roots, or zeros, of the denominator of the transfer function $H(s)$; they also can be found as the eigenvalues of the state matrix A from the state-variable description.

Roots of a polynomial are found with the **roots** command in MATLAB, and eigenvalues are found using the **eig** function. Let's consider the characteristic roots of the motor system. The characteristic polynomial is $s^2 + 11s + 10 + 5K_T$. Using $K_T = 0$, the roots are -10 and -1. Since we obtained these equations by assuming that the input and output signals had exponential decays e^{st}, then the response to an initial condition or impulse is the sum of two exponential decays, $(C_1 e^{-10t} + C_2 e^{-t})u(t)$.

There are four types of system responses, each governed by the root or roots closest to the imaginary, or $s = j\omega$, axis: underdamped, overdamped, critically damped, and marginally stable (or undamped). These roots are in general complex numbers of the form $s = \alpha + j\omega$ that result in system response terms having the form $Ae^{\alpha t}\cos(\omega t + \theta)$. Note that the real part of the root determines the rate of exponential decay, while the imaginary part—if it is not zero—determines the frequency of oscillation.

In the overdamped case, the root closest to the axis is purely real, and the response is dominated by an exponential term. In the critically damped case, the two roots closest to the imaginary axis are equal to each other and purely real; this response is a faster rising exponential, and the system has the fastest response time to any input. In the underdamped case, the system response has an oscillation that damps out over time. Finally, in marginal stability, there is no damping—only oscillation.

System Inputs and Initial Conditions

Simulating the operation of a continuous-time system with MATLAB requires specifying the inputs and initial conditions, in addition to characterizing the system via parameters of state-space or input-output representations. While **lsim** is the most general function for simulating system performance, as it allows any set of inputs and initial conditions, MATLAB also has some simpler functions that are useful for exploring the system response under some standard operations. If you are only interested in the zero-input response (the response to initial conditions only), then you can use the function **initial**. If you are interested in the step or impulse response, with zero initial conditions, then you can use the functions **step** and **impulse**, respectively.

Initial conditions are specified as a column vector. For the input-output representation, these initial conditions are specified in terms of the values of the output signal and its $(n-1)$st derivatives at the initial time. For the state-space representation, the initial conditions are the values of the states at the initial time. For the prosthetic limb, the state-space initial conditions would be the initial position, velocity, and acceleration of the limb, and the initial error and measurement values. For example, suppose the prosthetic limb is initially at rest on a stool and we want to see what happens if we take it off and let it drop. Using its height above the floor as a measure of the desired position, then the initial position will be the height of the stool, and the initial velocity and acceleration will both be zero. The error and measurement values will also be zero as the current position is the desired position at the initial time.

For the prosthetic limb, inputs are in the form of two vectors: one representing the desired position (or height of the prosthesis above the floor) at each time, and the other is a vector of times corresponding to each position (so that time is relevant). Consider the case where a patient wishes to walk at a smooth pace. You might use a sinusoid plus a unit-step function to represent the desired signal. Note that a sinusoid captures the "slowing down" when the limb is near its maximum height and on the ground, and the pace of walking is captured in the frequency of oscillation.

Root Locus

For the motor-tachometer subsystem, you can plot the characteristic root locations in the complex plane for each value of the tachometer gain K_T as it is increased from zero. The curve traced out in the complex plane using this approach is called a root locus curve, i.e. a locus or curve of characteristic roots. For systems in a standard feedback configuration, these curves can be automatically generated for you by MATLAB.

A general feedback control system has the form shown in Figure 5.4.2, where k is a variable gain. Having a feedback connection causes the overall system transfer function to change. If there is no feedback, the system has the transfer function $H(s)$. If there is feedback, the system transfer function is $\dfrac{Y(s)}{X(s)} = \dfrac{H(s)}{1 + kG(s)\,H(s)}$, known as the closed-loop response. The function $G(s)H(s)$ is known as the open-loop transfer function, which is the transfer function from the input to the output of the controller when feedback is disconnected.

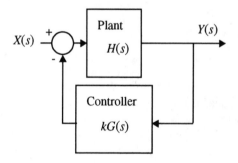

Figure 5.4.2 *A Feedback Control Configuration*

The denominator $[1+kG(s)H(s)]$ is the characteristic equation for this system; its roots define the system characteristic response. The closed-loop poles of the system depend upon k and the known pole and zero locations of $H(s)$ and $G(s)$.

Since the poles change with k, it's useful to know how they move and what k values will give useful pole locations. This information is concisely represented in a root locus diagram, which plots curves tracing the system poles for k increasing from zero. Root locus diagrams can be easily generated using MATLAB's **rlocus** function.

To use the **rlocus** function, use either the state-space representation found in Question 2 or determine the numerator and denominator polynomials of the open-loop transfer function and use them as input parameters to `rlocus`. The resulting root locus plot will be displayed automatically.

We are particularly interested in the values of k for which the system is stable, and what values of k yield critically damped, overdamped, and underdamped responses. When the poles lie on the imaginary axis, the system is operating at marginal stability. If any poles lie in the right half plane or if there are repeated poles on the imaginary axis, the system is unstable.

If the system characteristic equation does not have the form of $1 + kF(s) = 0$, then the `rlocus` function in MATLAB will not be useful. Instead, you must simulate the action of the root locus yourself by computing the roots for several values of k and plotting these points in the complex plane.

5.5 LABORATORY EXPERIMENT

The first patient scheduled to receive this experimental product is a world-class long distance runner who was in a severe car accident where her leg was severed just above the knee. She is willing to accept some small tremors when trying to move her leg slowly, but is adamant that she wants to be able to run as quickly as possible, as she plans to continue competing. Prior to attaching this prosthetic device, you are asked to simulate the operation of the artificial leg using the model in Figure 5.4.1 and to select reasonable parameter values. Specifically, you can adjust K_B, the gain/amplification of the myoelectric signal sensors, and K_T, the gain of the velocity sensor (a tachometer). You are told that clinical tests on this patient have indicated that the expected or average value of T is 0.25 s. With this information, you need to simulate the overall system response for three desired activities: the patient desires to pick up and hold her leg in a fixed position, to take her leg off a stool and place it on the ground, and to walk at various rates.

Some of the experienced engineers suggest that you initially (1) select K_T so that the motor-tachometer system is critically damped, and then (2) examine how the system characteristic roots and system response to a stepping up input vary for different K_B. They further suggest that you (3) determine the value of K_B where the system is marginally stable, as increasing the gain beyond this value will result in an unstable system. Based on their experience, these engineers suggest trying values of K_B in the range 0-30.[1]

They want you to pick values that result in a system which both meets the patient's needs and will ensure safe operation with minimal frustration. They are concerned that the clinical measurement of T is an inexact science; they would like the system to perform well for a range of T around 0.25 s—particularly lower values, as they think the delay might be less for a highly trained athlete.

---

[1] Actually, they did a Routh-Hurwitz test to determine this approximate range.

Fundamental Goals

Your objective is to select values of K_B and K_T that satisfy safety and patient requirements and are robust for small variations in T. These values will be the first values tried when the patient arrives for the first fitting of the prosthesis. You will need to provide guidelines telling the patient what type of performance she can expect from the device, given the gains you have selected. You can select values that meet the design requirements and determine the expected response of the overall system using simulation of the model shown in Figure 5.4.1. Be sure to verify the expected performance of your system for the following activities:

1.   Pick up leg and hold (or "stepping up"): a step response.
2.   Walk/run at different paces: a sinusoidal response.
3.   Take foot off stool and let fall to ground: an initial condition, zero-input response.

The senior engineers on the project have suggested the following tasks and observations as a reasonable approach to selecting the variable gains needed to achieve the desired results.

Problem 1.   Select K_T such that the motor/tachometer subsystem is critically damped. Examine the effect on the system poles and $y(t)$ of varying K_B in the range suggested, assuming a "stepping up" input. Generate a plot showing how the characteristic roots change as K_B is increased using the state-space representation found in Question 2 and **rlocus**.

> What can you say about the relative impact of different roots on the system response $y(t)$? Consider how their respective distances from the imaginary axis affects the system response. Which poles have the most influence on or "dominate" the system response?

Problem 2.   From your root locus diagram found in Problem 1, determine the value of K_B that corresponds to a marginally stable system. For this case, explore what happens to the output $y(t)$ if the pace of walking has the same frequency as that of the marginally stable characteristic root pair (i.e., the system resonant frequency).

> Indicate the value of K_B corresponds to a marginally stable system and provide a plot of the response to a sinusoidal input at the resonant frequency. What do you observe? Explain the system's behavior using the idea of resonance.

Problem 3.   Select values of K_B that result in underdamped, overdamped, and critically damped responses to a step input and identify the characteristic root locations. Explain how the system response depends upon the characteristic root locations. Consider such factors as response time and overshoot in physical terms, i.e., what happens to the patient.

> Compare your results to the marginally stable response. Which value of K_B allows the leg to move fastest, thereby allowing the fastest pace?

Enhancements

Problem 4.   Now see if you can achieve faster response without oscillation by allowing both K_B and K_T to vary (i.e., can you make the overall system critically damped). How much does the motor affect the system response? If the motor is represented by a gain of 1 (the ideal response), what value of K_B allows for the fastest leg motion?

> Attempt to move the pole or pole pair closest to the $j\omega$ axis as far from the $j\omega$ axis as possible. How does the distance of the pole from the imaginary axis impact system performance?

Problem 5.   After picking sensor and tachometer gain values that you consider to yield good performance in the overall system, simulate what might happen under different conditions that might occur during use in everyday life.

a. Check what happens if the patient is standing still with both feet on the ground but the motor has not yet come to a complete stop when the patient decides to lift her leg.

b. Consider how other activities or external forces might result in nonzero initial conditions in the overall system. Clearly, it is desirable to minimize unexpected movement of the prosthesis.

c. See how your system behaves if $T = 1/8$ s and $T = 1/2$ s. In these cases, might you be able to achieve better performance by using different values of the sensor and tachometer gains? Suggest whether you expect that you should increase or decrease these gains in each case and give a physical explanation.

d. See how fast the patient can run/walk. What happens if she tries to move faster than this?

> Compare the response of the system in each case to the results from Problem 3. Comment on the differences and explain how the pole locations have changed (if at all) and what effect the initial conditions have on the response.

Transform Domain Analysis

Problem 6.   Using MATLAB, determine the closed-loop and open-loop transfer functions from the state-space representations assuming that K_T is fixed at the critically damped value found in Problem 1 and $K_B = 2$. Verify that the eigenvalues of A are equal to the system poles.

> What are the open-loop and closed-loop transfer functions? What are the system poles?

Problem 7.   Using **rlocus**, generate the root locus plot. Identify values of K_B for critically damped, underdamped, overdamped, and marginally stable systems from the root locus plot with **rlocfind**.

> On your plot, indicate locations of the system operating points above. For each, label the value of K_B and the characteristic root locations.

Problem 8.   Using **freqs**, plot the system frequency response for each K_B value you found in Problem 7. Explain how the peaks in the frequency spectrum relate to the undamped natural frequencies of characteristic roots. What happens if your input is a sinusoid at these frequencies?

> Illustrate your understanding by identifying points for each K_B on your root locus plot and explaining how these impact the peaks in the corresponding frequency spectra.

Laboratory Assignment 6
Speech Modeling, Prediction, and Synthesis

PURPOSE

In this laboratory assignment, you will learn how to generate digitally synthesized speech by using a difference equation model for digital speech. Using MATLAB, you can easily implement digital filters defined by difference equations (for real-time speech synthesis, the digital filters are implemented on a dedicated DSP board). Using your model, you will explore the quality of speech you can digitally synthesize and the associated storage and transmission requirements. Using the Fourier transform, you will determine the bandwidth required for speech transmission, explore how speech frequency content changes with time, and compare the spectra of true and synthesized speech.

6.1 OBJECTIVES

By the end of this assignment, you should be able to

Time Domain

1. Use difference equations to model and synthesize speech.

2. Use MATLAB to simulate the time-domain response of difference equation models for DT LTI systems.

3. Compute transmission and storage requirements for speech technologies.

Transform Domain

4. Identify formant frequencies from sampled speech data and compute formant frequencies from linear prediction filter coefficients.

5. Determine the bandwidth and minimum sampling rate required for speech data.

6. Analyze effects of digital filtering using frequency-domain techniques and create inverse filters.

7. Compare spectra and signal to noise ratios (SNRs) for digital speech transmitted by quantized error signals and by quantized speech signals.

6.2 REFERENCES

Review topics

    1. Difference equation models for discrete-time (DT) systems

    2. Characteristic roots and responses for DT systems

Exploratory Topics

    1. Data windowing and windowing functions

    2. Frequency response of digital filters and resonant frequencies

    3. MATLAB's **filter** command

    4. Speech modeling and prediction for transmission and synthesis

Application Reference

    1. Rabiner and Schafer, *Digital Processing of Speech* (Englewood Cliffs: Pren-tice-Hall, 1978)

6.3 LABORATORY PREPARATION

Problems

Time Domain

Question 1.   Assume that you have a digitized speech signal that was sampled at 8 kHz. If the speech is broken down into 20 ms blocks, how many samples *NS* are there per block? If one second of a recorded speech signal is in a MATLAB vector, how many 20 ms blocks, *NBLKS*, are there (assuming no overlap of blocks)? What is a one-line MATLAB command that will extract the *n*th 20ms segment of speech, where $n = 1, ..., NBLKS$?

Question 2.   Suppose you wish to use as input to your speech model a train of equally spaced DT unit pulses, and would like the pitch to be 200 Hz. If the speech is assumed to be sampled at 8 kHz, how many DT samples are there per period, i.e., what is *N* in

$$x[n] = \sum_{i=0}^{NS-1} \delta[n - iN]$$

Question 3.   Look up the description of the MATLAB command **filter**. Determine how you need to define the vectors **a** and **b**, used as input to **filter**, in terms of α_i and G to create difference equations that will allow you to perform the following operations:

1. Give you $\hat{s}[n]$ as output when $s[n]$ is the input (linear prediction);

2. Give you $e[n]$ as output when $s[n]$ is input (prediction error);

3. Give you $\tilde{s}[n]$ as output when $e[n]$ is input (synthesis).

In each case, how can you ensure that you use the correct initial conditions for each successive speech block, defined as the final values from the previous block when using **filter**. Assume that you are in a loop that generates one 20 ms block at a time.

Question 4.   For the signal $x(t) = \sin(\omega t)$, where $\omega = 2\pi(100 + 50t)$, $0 < t < 1$ sec, plot a two-dimensional time / frequency representation of the signal. Put time on the horizontal axis and frequency on the vertical axis.

Question 5.   Given $e[n] = x[n] - a_1 x[n-1] - a_2 x[n-2]$,

(a) What is the transfer function for the linear prediction filter, where x is the input and e is the output?

(b) What is the transfer function of the inverse filter, where e is the input and x is the output?

(c) Given that $a_1 = 1.3789$ and $a_2 = -0.9506$, what are the formant frequencies in radians per second for the transfer function from part (b)?

(d) For the coefficients given in (c) above, sketch the frequency response. You might wish to read the text corresponding to Figure 10.4.3 in the Background section of Laboratory Assignment 10.

6.4 BACKGROUND

You work for a company that is developing a digital telephone answering machine for home computers. The system will sample data from a telephone line, detect rings, pick up the phone, speak a greeting, and record a message. The greeting will be a text message you type in, and the system will synthesize your speech from samples of your speech which will be recorded and analyzed when the system is installed. In addition, it compresses the recorded phone messages before saving to the hard drive so as to save space.

You are in charge of developing the compression and synthesizer portions of the system. As a first step, you attempt to model your own speech and determine how much compression is possible.

Speech Fundamentals

Physically, CT speech is produced when air from your lungs excites your vocal tract system. Sampling and quantizing CT speech results in digital speech. In telecommunications, speech is digitized by sampling at 8 kHz, using 8 bits per sample. The vocal tract behaves as a resonant cavity so that the signal emanating from your mouth is a weighted sum of delayed versions of the original vocal signal plus the excitations. We can model speech as a linear difference equation; the weights on the delayed signal versions are the coefficients of the model. Different sounds can be produced by using different inputs to and coefficients of this model.

Different types of speech sounds can be roughly categorized as either voiced or unvoiced, where the category is determined by the type of input used to produce the sound. Voiced sounds are produced by using a periodic sequence of pulses as input; the fundamental period of this sequence determines the resulting pitch. Vowels are voiced sounds; if you say "aah," you can feel the vibrations at the top of your vocal tract. Unvoiced sounds are produced by using random white noise as input (alone it sounds like static). These sounds generally are produced more by turbulent air flow in the mouth, such as "sh."

Discrete-Time Speech Models

A mathematical difference equation model for the vocal tract can be developed as follows. Since each successive DT speech sample is very closely related to previous samples, the value of the current speech sample can be estimated as a linear combination of previous samples.

$$\hat{s}[n] = \sum_{i=1}^{p} \alpha_i s[n-i]$$

$\hat{s}[n]$ is the estimate of the speech signal $s[n]$ for the nth sample. The error between the estimate and the original signal is

$$e[n] = s[n] - \hat{s}[n]$$

Prediction Model

Combining the two equations above yields a difference equation model of the prediction process for speech:

$$s[n] - \sum_{i=1}^{p} \alpha_i s[n-i] = e[n]$$

This prediction model is used in telecommunications to increase the number of voice signals that can be transmitted over a channel. If the coefficients α_i are known at both the transmitting and receiving ends, then only the error needs to be transmitted and the speech signal can be *reconstructed* at the receiving end using the difference equation above. At the transmitting end $s[n]$ is the prediction filter input and $e[n]$ is the filter output. It turns out that sending a sampled

---

error signal can result in substantial channel bandwidth savings; this idea is explored further in this laboratory assignment.

Synthesis Model

We can modify this same basic speech prediction model for use in speech synthesis. If our goal is to create a signal $\tilde{s}[n]$ that mimics the original sampled speech segment $s[n]$, then we can replace the error $e[n]$ by an input signal $x[n]$ multiplied by a gain G. Using the same form as the difference equation model for prediction results in the following difference equation model for speech synthesis:

$$\tilde{s}[n] - \sum_{i=1}^{p} \alpha_i \tilde{s}[n-i] = Gx[n]$$

If $Gx[n] = e[n]$, then the synthesized speech $\tilde{s}[n]$ should exactly match the original sampled speech segment $s[n]$; in this case the process is called reconstruction rather than synthesis.

Typically the coefficients α_i change every 10-20 msec as the vocal tract changes to produce different sounds. In synthesis, you apply a sequence of excitations to the model that has coefficients appropriate for that time interval to generate a sequence of sounds corresponding to a speech utterance.

System Characteristic Response and Roots

The characteristic response for a difference equation can be found from the characteristic roots or poles of the system, much in the same way that the characteristic roots for CT systems determine the system behavior. Here our DT development closely parallels that followed in Laboratory 4 for the CT case.

A response of the form z^n, where z is a complex number, can be shown to satisfy a linear, constant-coefficient difference equation with zero input. We can assume that the zero-input response is of this form, and $z^{-N}y[n]$, for $y[n] = z^n$, corresponds to a delayed version of $y[n]$, i.e., $y[n-N]$. Setting the input to zero and replacing each delay in the prediction or synthesis difference equations above by a power of z^{-1} results in the characteristic polynomial. The roots of the characteristic equation, given below, are the characteristic roots, which define the system characteristic response.

$$Q(z) = 1 - \sum_{i=1}^{p} \alpha_i z^{-i} = \prod_{i=0}^{p-1} (z - z_i)$$

Since the equation is order p, there are p characteristic roots z_i. For typical male speech, $p = 10$, and the roots form complex conjugate pairs so that all α_i are real valued.

Resonance and Formant Frequencies

Each complex conjugate pair of characteristic roots defines a damped sinusoidal characteristic response. Consider the root pair $z_i = |z_i| \exp(j\angle z_i)$; this root contributes a time-domain characteristic mode of the form $A|z_i|^n \cos(\angle z_i n + \theta)$. Note that the characteristic root magnitude determines the damping factor, and the phase of the characteristic root determines the frequency of oscillation, i.e. $\Omega = \angle z_i$.

For speech synthesis, these DT sinusoids represent sampled CT sinusoids. In this case, the CT frequency of oscillation is related to the DT frequency by $\Omega = \omega T$, where T is the sampling rate and ω is the CT frequency in radians per second. The corresponding frequency $f = \omega/2\pi$ Hz is called a formant frequency; it defines a resonant frequency of the vocal tract. Typical male speech ($p = 10$) is characterized by five formant frequencies. When the coefficients of the model change, different formant frequencies are produced, resulting in different intonations.

Analyzing and Synthesizing Speech

Synthesizing speech using the difference equation model requires that we first analyze a real speech segment to determine the best coefficients α_i for each 20 ms speech segment. Given these coefficients, we can synthesize speech by applying an appropriate input for each model. For voiced speech, a good model of the input source is a train of ideal impulses at a certain frequency (where the frequency determines the pitch). For unvoiced speech, a good model for the input signal is random or "white" noise. These source models rely on the impulse response of the linear prediction filter to generate the output sound.

The analysis procedure is as follows. First, the sampled speech signal is broken into 20-ms long sections using an analysis window. Windowing of data is discussed in a later section. Next, statistical analysis of the data, which determines how correlated adjacent sample values are, is used to determine the filter coefficients that provide the best speech prediction (i.e., that minimize the prediction error power)—see Rabiner and Schafer, *Digital Processing of Speech Signals*. For synthesis, the filter coefficients are used in the model and a periodic impulse train or white noise are used as inputs for each segment, generating synthesized speech.

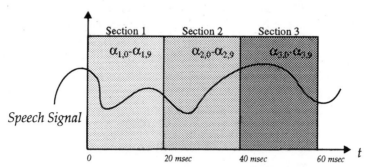

Figure 6.4.1 *Analysis Process on a Data Stream*

Transmission of Speech

A normal telephone line operates by sampling a person's speech, quantizing the samples to 8 bits, and transmitting these bits to the other end, where they are converted back into speech. An alternative method is to perform analysis and prediction as outlined previously, quantize the error signal, and transmit the resulting digital error signal and linear prediction coefficients.

Why do this? Normal speech requires 8 bits x 8 kHz = 64,000 bits per second to be transmitted. Suppose that the error signal can be quantized with 4 bits instead of 8 and that each coefficient can be represented with 16 bits. Then to transmit an equivalent amount of information, 4 bits x 8 kHz + 16 x 10 coefficients x 100 ten ms chunks per second = 48,000 bits per second—75% of the previous rate. If only 1 quantization bit is necessary for the error signal, 24,000 bits per second are necessary—37.5% of the previous rate. Using this technique, two people can have conversations in the same space as one person, a tremendous savings.

Both quantizing the sampled speech, as in the first paragraph, and quantizing the prediction error, as above, add distortion to the reconstructed speech. In some of the lab experiments, you will observe the differences between original speech segments and their reconstructed versions resulting from using different numbers of quantization levels. Representing the coefficients with 16 data bits also introduces some quantization error, which can lead to poor quality reproduction on the receiving end. This property is explored in the analysis questions.

In the computerized answering machine, the number of bits to be stored directly relates to the amount of space required. Since the library of sounds necessary to reproduce the answering machine greetings can become very large very quickly, having a good compression technique will allow more message flexibility and require less memory.

Windowing a Data Stream

As discussed above, the α_i coefficients change every 10-20 ms. For every 10 ms block of speech a new set of α coefficients must be calculated from the sampled speech data. The process of extracting a 10 ms block of speech from the entire segment is called windowing.

The simplest type of windowing involves taking the speech samples in the current 10 ms segment as data. This operation is mathematically equivalent to multiplying the entire signal by a rectangular function having a value of 1 in the region of interest and 0 everywhere else, just as when you multiply a signal by a difference of time-shifted unit step functions. This window function is called a rectangular window. At the edges of the data region, there is a sharp transition from signal to nothing, which can cause problems in analysis.

A better way to window the data sequence is to multiply by a function that has a smooth transition from one end to the other. The most common function that does this is called the Hamming window, which can be calculated using MATLAB for any length by the function **hamming**.

To understand why the Hamming window is preferred to the simpler rectangular window, it is instructive to look at the impact of windowing in the frequency domain. Since windowing a signal is a multiplication operation in the time domain, it corresponds to convolving the Fourier transform of the window function with the frequency spectrum of the speech segment. If the window transform approximates an impulse in frequency, then this convolution operation yields a frequency spectrum identical to the original speech spectrum. However, the less the window transform is like an impulse, the more windowing distorts the original speech signal spectrum. In Figure 6.4.3, the Fourier transforms of a rectangular window and a Hamming window are compared. Note that the Hamming window has more drop-off before flattening out, called stopband attenuation, but its main lobe is twice as wide as that of a rectangular window.

Figure 6.4.2 shows both Hamming and rectangular window functions.

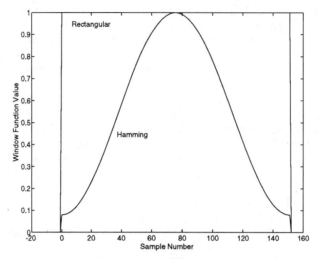

Figure 6.4.2 *Hamming and Rectangular Windows*

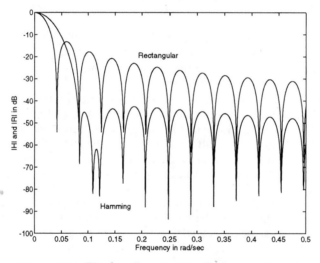

Figure 6.4.3 *Window Functions in Frequency Domain*

Implementing Difference Equation Models in MATLAB

The **filter** command in MATLAB can be used to compute the response of a difference equation model to specified input signals and initial conditions. Using difference equations to operate on an input signal is called filtering. The initial conditions of the various delay elements play an important part in the output of a difference equation. For example, computing $y[n]$ recursively using the difference equation $y[n] = x[n] + x[n-1] - y[n-1]$ requires knowledge of $y[n-1]$ at each time. The initial value of $y[n-1]$ is the value of the output prior to application of a new input; different initial values result in different initial responses to the input. Often, when using difference equations for continual filtering of a constantly applied input, the initial conditions are assumed to be zero, as the response to these initial conditions will only affect the filter response at system start-up.

For this experiment, however, the initial conditions are very important. Since speech is a continuous phenomenon but we are breaking it into 20 ms chunks, we would like the filter output at a 20 ms boundary to be consistent with the values from the previous block. Otherwise, there will be "pops"—caused by the errors in the initial conditions—in the output. The last output samples in the previous 20 ms segment should be used as initial conditions for the current 20 ms segment.

6.5 LABORATORY EXPERIMENT

Voiced Speech Models

Your objective is to generate a model for your speech, and then to try synthesizing your speech. As a first pass, you want to try to synthesize a purely voiced segment of speech.

Problem 1. Record yourself saying "We were away a year ago." Sample at 8000 samples/s and store the resulting signal in a MATLAB vector. To save time, you may want to process only a portion of this sentence when you are troubleshooting.

Problem 2. We have provided you with a MATLAB function file **P_6.m**. Given your original sampled speech vector, an excitation vector, and the number of full 20 ms blocks in your speech segment, **P_6.m** generates the prediction error, predicted speech, speech reconstructed using the prediction error as input, and speech synthesized using the excitation vector as input. Information regarding how to use this function is included as comments in the file. Variable definitions can be found by typing **help P_6.m**.

You need to edit **P_6.m** and add in the commands to create the appropriate a and b vectors, segmented original speech, excitation vectors, and filter command to generate (a) $e[n]$, (b) synthesized speech $\tilde{s}[n]$, and (c) speech synthesized using $x[n]$ as input. The use of the filter command is illustrated in the code by generating $\hat{s}[n]$ from $s[n]$. You may wish to review your answers to Question 1 and 3 prior to coding.

Problem 3. Generate a DT signal that is a periodic sequence of DT unit impulses. Pick your period to generate a pitch somewhere in the range of 50-300 Hz. Use your answer to Question 2 for guidance.

Problem 4. Use **P_6.m** to generate synthesized speech using the impulse train. Look at the resulting plots of and listen to the signals generated by **P_6.m**.

Improvements on the Voiced Model

Some ideas to consider and possible ways to improve your synthesized speech are described below. Use these suggestions to further explore speech synthesis using difference equations.

Problem 5.   Try different pitches for your voiced excitation in Problem 3. You may want to try to estimate a reasonable pitch period from your original speech vector by looking for the significant periodicity present in the time- or frequency-domain plots of each speech segment.

Problem 6.   Try synthesizing a different segment of speech, such as "Sally sells sea shells by the sea shore," using the same approach as above. Is it intelligible? Does it retain the same perceptual characteristics as the original speech segment? Pay particular attention to the "sh" sound.

Problem 7.   Try using an unvoiced excitation vector as input for the speech segment in Problem 6 by using the MATLAB command **randn** to generate white Gaussian noise having zero mean and unit variance (the default for **randn**). How does using this input instead of periodic impulses impact the perceived quality of the synthesized speech segments for the sentences from Problems 1 and 6?

Problem 8.   Using what you've learned in Problems 5 through 7, try creating an excitation vector which uses both unvoiced and variable-pitch voiced inputs to create more realistic-sounding speech. Try to do this for an arbitrary speech segment, as well as for the sentences above. You may want to analyze $e[n]$ to determine whether voiced or unvoiced excitation is appropriate for synthesizing a given segment of speech. For segments where $e[n]$ looks more random, use an unvoiced excitation, and where it looks more periodic, use a voiced excitation.

Transmission of Speech

Problem 9.   Obtain an error signal for a speech sequence using **P_6.m**. Normalize the error signal so that the maximum value is 1.0 and the minimum value is −1.0. Quantize the resulting error signal using 5 bits by using **P_6_9.m**. Plot the normalized error signal and the quantized error signal at the same time, and note any differences. Also plot the quantized error signal subtracted from the normalized error signal. What is the maximum value of the difference? Repeat for 4 bits, 3 bits, 2 bits, and 1 bit. How does the number of quantization levels affect the accuracy of your quantized representation of $e[n]$?

Problem 10.   Resynthesize the speech waveform using quantized error signals for the E1 input. Try using 4 bits, 2 bits, and 1 bit for quantization. Listen to the sampled speech and resynthesized speech, and comment on the differences and the relative rate of improvement observed as you increase the number of bits. Remember to undo your normalization of the error signal so that the volume of the synthesized speech will be the same.

Problem 11.   Calculate how many bits per second are needed to transmit the speech signal using 5-bit, 4-bit, 3-bit, 2-bit, and 1-bit quantization of the error signal and 16-bit quantization for each filter coefficient. If you want to satisfy the conflicting goals of minimizing the bit rate and maximizing the quality of reproduction, which quantization level (using your results from Problem 10) provides the best trade-off?

> Compare the bit rate for sampled speech with that of a system which transmits only the quantized prediction error. Try to quantify the bit rate vs. quality trade-offs and suggest potential applications.

Frequency Domain Analysis of Speech

Problem 12.   Record yourself saying "Ick Ack" at 44.1 kHz. Perform a Fourier transform of the first 8192 sampled data points using **fft**, and plot the magnitude of the frequency spectrum. Find the highest frequency represented, and determine the maximum sampling rate needed to accurately represent the data.

> How much information is lost if the sampling rate is 8 kHz, the telephone standard?

Problem 13.   Speech is the output of a time-varying system with time-varying inputs. You are to perform a time-frequency analysis of your sampled speech using the MATLAB function **specgram**. From your time-frequency plot, identify the three major formants in your speech segment and track how they change over time.

Problem 14.   Obtain a set of coefficients for one speech segment. Find the frequency response of the synthesis filter for these coefficients, and compare it to the frequency spectrum of the speech data for that segment. What similarities and differences do you notice?

How well does the linear prediction filter model the frequency content of the speech?

Analysis Questions

Problem 15.   *Coefficient Variation Between Blocks:* Determine the model coefficients, characteristic roots, and formant frequencies for some neighboring speech segments and for some different sounds. Supply a table of these values below and attach your calculations (which can be done via MATLAB). How much variation do you observe in these from segment to segment and for different sounds?

Problem 16.   *Finite-Precision Effects:* For the segments you used in Problem 15, what can you say about the stability of the difference equation for each segment? Now try reducing the number of decimal places you use to represent each coefficient. How do the characteristic roots, formant frequencies, and system stability of each segment change, if at all? What are the implications for digital storage of the model coefficients?

Problem 17.   *Storage and Compression Implications:* For the original voiced speech segment that you recorded, determine the storage savings achievable by storing only the coefficients, rather than the sampled speech segment. Be sure to use as many bits per coefficient as needed to ensure that the model is not compromised. How does your answer change if the speech was digitized as high quality audio (44.1 kHz sampling rate and 16 bits per sample)?

Laboratory Assignment 7

Image Deblurring and Decoding

PURPOSE

In this laboratory assignment, you will learn how to express DT convolution in a matrix/vector format and how to use this format for deblurring and decoding images. You also will learn how easy it is to display and process images using MATLAB. You will see that using frequency-domain techniques simplifies the process of determining the length of the blur and provides insight into the effects of blurring and deblurring on the image spectrum.

7.1 OBJECTIVES

By the end of this assignment, you should be able to:

Time Domain

1. Manipulate and process images in MATLAB.

2. Express convolution in matrix form.

3. Use matrix convolution for deblurring and decoding images and for system identification.

Frequency Domain

4. Use cepstral processing to identify the length of a rectangular, one-dimensional blur.

5. Predict the effect of image blurring and deblurring on the frequency spectrum of an image.

7.2 REFERENCES

Review Topics (italicized items are required only to perform frequency-domain exercises)

1. Discrete-time convolution

2. Putting a system of equations in matrix form

3. Matrix algebra

4. *Fourier analysis*

Exploratory Topics
      1. DT convolution as a matrix multiplication

      2. Representing linear blurs as a matrix

      3. Digital image representations

      4. Image deblurring

      5. Separable, linear predictive image encoding

      6. *Cepstral processing*

      7. *Two-dimensional frequency analysis*

Application Reference
      1. W. K. Pratt, *Digital Image Processing*. Prentice-Hall, 1985.

7.3 LABORATORY PREPARATION

■

Problems

Question 1.   Assume that you have a 3 x 3 image Y that corresponds to an original image X which has been corrupted by a two-point blur ($N = 2$). Determine the elements and size of the matrix H as defined in the background deblurring section. Starting from the definition of DT convolution, show that the distorted image Y for this 3 x 3 example is $Y^T = HX^T$.

Question 2.   Show how you can use the MATLAB command **toeplitz** to easily generate the blur matrix H from the unit impulse response vector h.

Question 3.   Again assume that your original image X is 3 x 3. Furthermore, assume that the image X was encoded using $p = q = 1$ to obtain the encoded image E. Using the definition of DT convolution and/or your results from Problem 1 above, show that the 3 x 3 encoded image E is related to the original image by $E = H_\beta X H_\alpha^T$ as defined and discussed in the image decoding background section. Show the size and elements of H_β and H_α in terms of α_1 and β_1.

Question 4.   How do the last two equations in the image decoding background section need to be changed if the prediction error along the columns is generated before the prediction error along the rows?

Question 5.   Explore how images are displayed and represented as matrices of integers in MATLAB. Color information is coded in a colormap (a lookup table of sorts). Images can be stored in **image_name.mat** files and loaded into MATLAB by entering **load image_name**. MATLAB has several standard images available. Go through the following tutorial prior to beginning the lab (just turn in a diary of your commands):

• Enter **load clown**

• Now check your variables using **whos**. You should have three variables: **caption** (stores image caption information), **map** (stores color information), and **X** (stores image intensity information).

• Look at some of the elements in **X** (e.g., type **X(65:75,100:110)**). The values in the array should be integers.

• To display the image, enter **image(X)**. You should now see a clown in the figure window, although the color may appear unnatural.

• To get the correct colors, enter **colormap(map)**. Now the clown should appear to have the correct colors.

After you perform mathematical operations on an image matrix X, the elements may no longer be integers. To convert them back to integers prior to display, simply use **image(round(X))**.

To "brighten" an image, enter **image(K*X)**, where K is some integer greater than one, or use **brighten**. We also suggest that you refresh your memory on the **inv** command for inverting matrices.

Question 6.   A two-point horizontal image blur can be represented by the difference equation $y[n] = 1/2(x[n] + x[n-1])$, where $x[n]$ denotes pixel values from a row of the original image and $y[n]$ denotes the corresponding pixel values from the same row in the blurred image.

(a) What is the transfer function for the blur? If you have not covered DT transfer functions in class, you may wish to read the discussion in Laboratory Assignment 6 on Characteristic Roots and Responses.

(b) Sketch the frequency response of the blur. If you are not familiar with how to predict the DT frequency response from the filter pole locations, you might want to jump ahead and read the discussion in Laboratory Assignment 10 corresponding to Figure 10.4.3. What type of filter is this?

(c) Sketch the frequency response of an inverse filter that would remove the blur. What type of filter is this? Why might this filter emphasize noise?

Question 7.   You have a 3 x 3 image, expressed as a matrix $x[i,j]$, $i = 0, 1, 2, j = 0, 1, 2$, where the elements of the matrix are $x[i,j] = 1$ for $i = j$, and 0 for $i \neq j$. Determine the 3 x 3 two-dimensional DFT of $x(i,j)$ and call it $X[k,l]$. Given the properties of 1-D Fourier transforms, does your result make sense? Think about taking 1-D transforms along rows and columns.

Question 8.   Assume that a blur function $h(t)$ of length T is convolved with an arbitrary signal $x(t)$ such that $y(t) = h(t) * x(t)$.

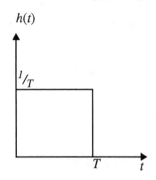

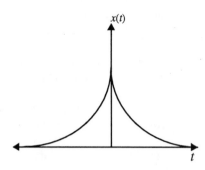

(a) Sketch the magnitude and phase of the frequency spectrum for $x(t)$ and the frequency response corresponding to $h(t)$. At what frequencies does $|H(j\omega)|$ become zero?

(b) Sketch the magnitude of the frequency spectrum $|Y(j\omega)|$ resulting from the convolution of $h(t)$ with $x(t)$.

(c) Sketch the spectrum resulting from $\log(|Y(j\omega)|)$. How can you find T from this plot?

(d) What happens to $\log(|Y(j\omega)|$ if the input $x(t)$ is a rectangular function of length T_1? How might this impact your ability to determine the length of the original blurring function from this plot?

Question 9.   Consider the unit pulse response of a simple 1-D rectangular blur, $h[n] = 1$ for $n = 0, 1, ..., L - 1$, and 0 otherwise. If you compute the length $N = 2^r > L$ or $N = L$ DFT, determine where zeros will occur for the resulting $\sin(x)/x$ frequency response. With this information, how might you use the DFT to compute the length of the blur from a blurred signal? A word of caution: the zero crossings of $\sin(x)/x$ may not occur exactly at a DFT sample of the spectrum. How can you estimate where these occur?

7.4 BACKGROUND

■

Scenario

Professor Abducted   *{excerpted from local news}*

A University professor was kidnapped yesterday in broad daylight. Witnesses say that as the professor was leaving The Bank, a mysterious looking car pulled up to the curb. Two men wearing ski masks hopped out and stuffed the professor into the back seat. The green car then quickly turned off of Main Street. Police have decided this was probably not a gang-related kidnapping.

The police may be aided in their search for this wealthy professor by a surveillance camera that was mounted on the outside of The Bank. The camera obtained a horizontal sweep of the green car that was reported. Unfortunately, the picture was blurred because of the motion of the camera. The police are attempting to get help in deblurring this picture from specialists.

Your Task

As indicated in the news report, the police have asked for help in deblurring an image. They have stored the image digitally on a computer, so advanced image processing techniques can be used to determine the license plate number. You have been given the task of deblurring the license plate image and have been instructed to be on call for any further technical tasks. It turns out that your assumptions about the type of blur and its impact on the image are correct. The police are able to trace the car to a rental agency. Unfortunately, the car was paid for in cash and a false name was used.

The police did discover a CD-ROM containing some image files under one of the seats. One image is that of the professor who was abducted. All but one of the rest of the images are encoded so that they're unrecognizable. The police give you a second task: to identify the code used and decode the images. From the decoded images, they hope to find out which, if any, additional professors might be in danger so that they can be notified and put under police protection. You also are to note if any of the pictures appear to have been digitally "doctored." Fortunately your job is made easier as one decoded image appears to correspond to two of the encoded images. You anticipate that this should help you to identify the code used by the perpetrator in encoding the images so that you can decode the images and identify possible victims.

Digital Image Representation

Digital images are made up of many "dots," or pixels (picture elements). By placing these pixels close enough to each other, the images viewed on a computer display or hardcopy appear to be continuous. On a digital computer, the brightness and color information for each pixel is encoded by a number in an array, or, equivalently, an element in a matrix. The location of each value within the array is indexed by two integers, e.g., $X[3,4]$ identifies the pixel value located in the matrix X in the third row, fourth column.

Usually the values in the arrays are integers from 0 to 2^n-1, where n is the number of bits used to represent the brightness of each pixel. For example, consider a black-and-white picture where the luminance, or brightness, of each pixel is stored using 8 bits. In this case, the relative brightness of each pixel can be represented as one of 256 possible levels, called gray levels. Usually black is encoded by a value of 0 and white by a value of 255.

The color information for pixels in the image also is usually encoded as integers stored in matrix arrays. In MATLAB, images are stored as integers in arrays and the color information—

---

how each pixel value maps to a certain color—is stored separately. The use of MATLAB for storing, processing, and displaying images is discussed in Lab Prep Question 6.

Modeling Linear Blurs for Image Deblurring

A horizontal motion blur represents a linear system. Horizontal blurring causes each pixel in an image to contain some information from N previous pixels of the same row. A simple model of horizontal blurring using DT convolution is

$$y[l, n] = \sum_{k = max(0, n - N + 1)}^{n} x[l, k] h[n - k]$$

where $h[n]$ is the unit pulse response,

$$h[n] = \frac{1}{N} \qquad n = 0, \dots, N-1$$

$x[m,n]$ and $y[m,n]$ are arrays of numerical values that represent pixel values in the original and blurred pictures, respectively. Note that this "blurring" operation corresponds to replacing a pixel value by an average of the surrounding pixel values. The first index denotes the mth row, the second index indicates the nth column, and the image is $L \times L$. For each row l, the DT convolution above can be expressed in matrix form as $y_l = Hx_l$.

$$
\begin{bmatrix} y[l, 0] \\ y[l, 1] \\ \dots \\ y[l, L-1] \end{bmatrix}
=
\begin{bmatrix}
h[0] & 0 & \dots & \dots & \dots & \dots & \dots & 0 \\
h[1] & h[0] & 0 & \dots & \dots & \dots & \dots & 0 \\
\dots & \dots & h[0] & 0 & \dots & \dots & \dots & 0 \\
h[N-1] & h[N-2] & \dots & h[0] & 0 & \dots & \dots & 0 \\
0 & h[N-1] & \dots & \dots & h[0] & 0 & \dots & \dots \\
\dots & \dots & \dots & \dots & \dots & \dots & \dots & 0 \\
0 & \dots & \dots & 0 \, h[N-1] & \dots & h[1] & h[0]
\end{bmatrix}
\begin{bmatrix} x[l, 0] \\ x[l, 1] \\ \dots \\ x[l, L-1] \end{bmatrix}
$$

If you transpose both sides of the equation above for each row and then form image matrices from the x's and y's, the blurring operation can be represented by

$$
\begin{bmatrix} y[0, 0] & \dots & y[0, L-1] \\ \dots & \dots & \dots \\ y[L-1, 0] & \dots & y[L-1, L-1] \end{bmatrix}
=
\begin{bmatrix} x[0, 0] & \dots & x[0, L-1] \\ \dots & \dots & \dots \\ x[L-1, 0] & \dots & x[L-1, L-1] \end{bmatrix}
\begin{bmatrix} H \end{bmatrix}^T
$$

yielding $Y = XH^T$. Since H is square, deblurring can be accomplished by simply inverting the matrix H^T to solve for $X = Y(H^T)^{-1}$. Note that the matrix ordering is important.

Decoding Linear Prediction Error Images

If you "brighten" the encoded images, you notice that some picture structure exists. While many edges are evident, horizontal and vertical edges seem less obvious, and the image appears noisy. Based on this information, you deduce that the perpetrator used separate horizontal and vertical linear prediction filters.

Recall from your speech processing experience that a linear prediction filter generates the error between a sample (pixel) value and its predicted value as estimated from a linear combination of previous sample (pixel) values. You decide that it is likely that the image X was encoded by generating an error matrix using a two-step linear prediction process.

You assume that the perpetrator first computed the prediction error along each row:

$$e_I[l, n] = x[l, n] - \hat{x}[l, n] = x[l, n] - \sum_{k=1}^{p} \alpha_k x[l, n-k]$$

yielding an intermediate error matrix E_I. Note that if you know the "code," or prediction coefficients, α_k, you can invert this process using the deblurring approach outlined above. Similarly, in the second stage, the prediction error along each column in E_I was generated using prediction coefficients β_k, yielding the elements in the total error matrix E.

$$e[l, n] = e_I[l, n] - \hat{e}_I[l, n] = e_I[l, n] - \sum_{k=1}^{q} \beta_k e_I[l-k, n]$$

Again, this process can be inverted if the "code," or values of β_k, is known.

Using the same approach as in the previous section, we can define the unit pulse response vectors for the row and column encoding processes, respectively, as

$$h_\alpha^T = \begin{bmatrix} 1 & -\alpha_1 & \ldots & -\alpha_p \end{bmatrix}$$

$$h_\beta^T = \begin{bmatrix} 1 & -\beta_1 & \ldots & -\beta_q \end{bmatrix}$$

Using the same approach as in the image deblurring section, we can express the image encoding process mathematically as two matrix multiplications:

$$\begin{bmatrix} e[0,0] & \ldots & e[0, L-1] \\ \ldots & \ldots & \ldots \\ e[L-1, 0] & \ldots & e[L-1, L-1] \end{bmatrix} = \begin{bmatrix} H_\beta \end{bmatrix} \begin{bmatrix} x[0,0] & \ldots & x[0, L-1] \\ \ldots & \ldots & \ldots \\ x[L-1, 0] & \ldots & x[L-1, L-1] \end{bmatrix} \begin{bmatrix} H_\alpha^T \end{bmatrix}$$

where the H matrices are defined in terms of the unit pulse response vectors h in the same way as for the blur matrix. If the α_i's and β_j's are known, then the original image can be obtained from the encoded image by inverting the unit pulse response matrices:

$$X = \left[H_\beta\right]^{-1} E \left[H_\alpha^T\right]^{-1}$$

If all the matrices resulting from one complete encoding sequence—including the intermediate error matrix E_I in addition to the original and final encoded images, X and E, respectively—are known, then the α_i's and β_j's can be determined from

$$H_\alpha^T = X^{-1} E_I$$

$$H_\beta = E E_I^{-1}$$

Again, note that matrix ordering is critical. Note that if the image is encoded by generating the prediction error along each column first, then these equations need to be modified.

Frequency-Domain Concepts for Determining Blur Length (Cepstral Analysis)

In the first part of this experiment, you determined the blur length by trial and error, assuming a linear blur. You surmised that the value of N that yielded the best deblurred image is assumed to be the length of the original blur. It would require considerably less time if you were able to use a more structured analytical approach to determine this length. An understanding of frequency-domain concepts provides the fundamental tools needed to estimate the blur and results in a technique, outlined below, called cepstral analysis.

The basic idea is the following: consider a linear blur of length π as shown below.

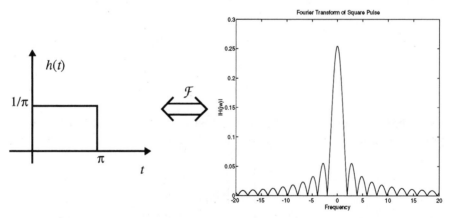

Figure 7.4.1 *Fourier Transform of a Rectangular Pulse*

From Fourier transform theory, you should recognize that the magnitude of a linear blur has the standard |sin(x)/x| form. You can determine the exact spectrum from any standard Fourier transform table. Note that the spacing of the zeros (points where the function is equal to zero and thus touches the horizontal axis) is related to the length of the blur. This information is very useful in determining the blur length and is used in cepstral analysis, a technique for using simple mathematical manipulations to highlight the positions of these periodic zeros.

Assume that the original unblurred signal has the spectrum shown below:

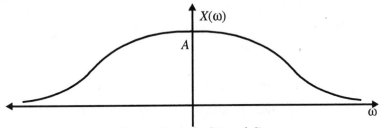

Figure 7.4.2 *Sample Signal Spectrum*

Then the blurred image magnitude spectrum $Y(\omega)$ will be the product of the blur frequency response $H(\omega)$ and the signal spectrum $X(\omega)$, i.e., $Y(\omega) = H(\omega)X(\omega)$, resulting in

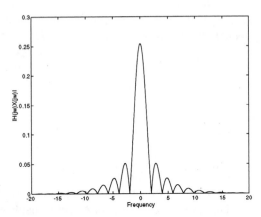

Figure 7.4.3 *Signal Spectrum of Blur*

Note that $Y(\omega)$ still has the same zero spacing as $H(\omega)$! Since we have the blurred signal available to us, we can look for the frequency spacing ($\Delta\omega$) between periodic zeros in the magnitude of the Fourier transform. The length of the blur is then $T = 2\pi/\Delta\omega = 2\pi/2 = \pi$.

Difficulties can occur if the original signal $x(t)$ has spectral zeros or is corrupted by noise. Noise essentially adds a constant value at all frequencies, so there will be no zeros. Zeros or dips in the magnitude spectrum of $x(t)$ may cause some confusion as to which dips arose from the blur and which dips came from the original image.

The solution is twofold:

1. Emphasize spectral dips using logarithmic functions so that spectral dips yield a plot with large peaks; this is called the cepstrum.

2. Look for periodic peaks in the cepstrum.

The procedure is as follows: given the blurred image **Y**,

1. Take one row $y_i(t) =$ **Y(i, :)** from the image.

2. Compute $|Y_i(\omega)|$ using the DFT (**fft**).

3. Compute $-\log(|Y_i(\omega)|)$. At frequencies where $|Y_i(\omega)|$ is very small, this function will result in a very large positive number.

4. Clean up your signal by setting all negative values to zero.

5. Look for periodic peaks and use this information to determine the blur.

A word of caution: determining spectral nulls using digital Fourier analysis via the DFT, as in MATLAB, can cause some problems. Specifically, a spectral zero may fall in between spectral samples and you will have to estimate where these nulls occur. One way to minimize this problem is to zero-pad your vector by designating a length for the **fft** function in MATLAB that is longer than your original signal vector length. It is advisable to designate a length that is a power of 2 as this substantially speeds up the computations. Ascertaining T using the method above requires that you express the frequency spacing between periodic nulls as $\Delta\omega$; note that this requires that you convert from the integer index to radians per second. It turns out that you can determine the blur length directly from $\Delta\omega$ and the length of the DFT.

7.5 LABORATORY EXPERIMENT

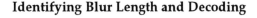

Identifying Blur Length and Decoding

You need to figure out how many of the previous pixel values are affecting each pixel in the same row of the image (i.e., find N). The value of N will be between 1 and 20. When you can best identify the license plate number, you have found the correct N. You might need to brighten the picture, enhance the contrast, make the image window bigger, or use some edge detection filters to make sure your number is correct. The blurred image is stored as **blur.mat**.

Problem 1.   Using the results from the Background section and Laboratory Preparation problems, determine N experimentally by trial and error; write an M-file to deblur the image for different values of N. Identify the license plate number and length of the blur.

Problem 2.   Extract out a row of the blurred image surrounding the license plate of the car. Using cepstral techniques, determine the blur length and deblur the image. Which method of determining N, iteration or cepstral analysis, is easier?

Problem 3.   Plot 2-D frequency spectra of the blurred and deblurred images. Determine the frequency response of the blurring and deblurring filters.

Identifying the Linear Code and Decoding

Recall that the CD-ROM found under the car seat contains some images. The CD contains an image of the professor who was abducted, a set of encoded images, and one set of images that appears to be an original, an image resulting from the first encoding stage of this original, and the final encoded image. You are charged with determining the code, decoding the images, identifying who might be in danger, and flagging any pictures that appear to have been digitally "doctored."

The set of encoded and decoded image pairs to be used for identifying the code is stored in **encset.mat**. The variables **G**, **Genc1**, and **Genc2** correspond to the original image, the image after encoding in one direction, and the image after encoding in the other direction. Use these images to determine p, q, α_i, and β_j for all i and j. Assume that all images were encoded using the same parameters; you can therefore use these same values to decode the remaining images **[A-H]enc**. Note that some of the remaining images may be of different sizes, and thus you will need to construct the correct size H_β and H_α for each picture. You may want to write an M-file to construct these given the size and the coefficients of the unit pulse response vector using the **toeplitz** command.

Problem 4.   Identify the terms of the unit pulse responses used to encode the rows and columns. What conditions must X, E_l, and E satisfy for you to be able to identify these unit pulse responses? Consider what happens if the images are not square and suggest a solution.

Problem 5.   Decode the images and identify the pictures. What conditions must the matrices, formed from the row and column unit pulse responses, satisfy for you to be able to decode the images? Are these matrices always square?

Analysis Questions

Problem 6.   How does your procedure for identifying and decoding the images change if the prediction error along each column is generated first, followed by the prediction error
along each row?

Problem 7.   Now assume that the prediction error matrix is formed by prediction from the rows and
columns simultaneously, i.e., $e[l, n] = x[l, n] - \hat{x}[l, n]$ and

$$\hat{x}[l, n] = \alpha_1 x[l, n-1] + \alpha_2 x[l-1, n-1] + \alpha_3 x[l-1, n].$$

Assume that X and E are 3 x 3 matrices from which the vectors x and e, respectively,
are formed by stacking columns of the matrices on top of each other. Determine the elements of the matrix H, where $e = Hx$, in terms of α_i, $i = 1, 2, 3$. If you are given H and
e, can you determine x?

Laboratory Assignment 8
Digital Sonar for Localization and Signaling

PURPOSE

In digital communication systems, information is encoded and transmitted as sequences of binary ones and zeros. These ones and zeros are represented by continuous-time waveforms for transmission. During transmission, these waveforms are often corrupted by noise. For accurate communication of the binary information, the receiver must be able to correctly identify the transmitted sequence of ones and zeros.

In this laboratory assignment, you will learn how to design the impulse response, or, equivalently, the frequency response, of a CT LTI system so that the receiver can best estimate the original signal even in the presence of substantial corrupting noise. Such systems are used by receivers in digital communication systems and are called matched filters. You will design matched filters to reduce noise in a digital sonar system so that you can correctly identify the distance to a friendly ship and accurately decode a message that it transmits. You will also explore how the Fast Fourier Transform can be used to implement a computationally efficient matched filter.

8.1 OBJECTIVES

By the end of this assignment, you should be able to

Time Domain

1. Identify basic building blocks in digital communication systems.

2. Design a matched filter by specifying the unit impulse response needed to implement a correlation receiver.

3. Implement a matched filter using continuous-time convolution.

4. Use your matched filter implementation for estimating sonar signals in the presence of noise and distances of objects using active sonar.

Frequency Domain

5. Design a matched filter by specifying its frequency response.

6. Implement a computationally efficient matched filter by using the radix-2 FFT to perform convolution.

7. Explain how a matched filter works in both the time and frequency domains.

8.2 REFERENCES

Review topics

      1. Continuous-time convolution

      2. Fourier transform properties

      3. System frequency response

Exploratory topics

      1. Correlation receivers

      2. Matched filtering

      3. Computationally efficient convolution using the FFT

      4. Sonar digital communication systems

8.3 LABORATORY PREPARATION

Problems

Question 1.   Assume that logical ones are represented by shifted positive rectangular pulses, $p(t) = u(t) - u(t - T/2)$, and logical zeroes are represented by shifted negative rectangular pulses, $-p(t)$. Assume that the bit rate is T bits per second.

(a) What is the unit impulse response $h(t)$ of a causal, matched filter for this system?

(b) Compute $h(t) * p(t)$ and $h(t) * [-p(t)]$.

(c) If $x(t)$ is the waveform representing a sequence of alternating ones and zeros starting with a 1 at $t = 0$, sketch $x(t)$ and $y(t) = h(t) * x(t)$. If you sample $y(t)$ at T-second intervals, determine at what times you should sample, and find a criterion for determining whether a one or a zero was sent using these samples.

Question 2.   Assume that you sent out a sonar test signal $s(t) = u(t) - 2u(t - T/2) + u(t - T)$ and receive the signal $s(t - T_d)$, where $T_d \gg T$.

(a) define $h(t) = s(T - t)$ and sketch $y(t) = s(t - T_d) * h(t)$. Assume $T = 0.1$ and $T_d = 20$.

(b) From $y(t)$, how can you determine the time delay T_d?

Question 3.   Look up the MATLAB function **conv**. Show how to use this function to generate a signal $y(t) = x(t) * h(t)$. If $x(t)$ is represented in MATLAB by a vector of length M and $h(t)$ is represented by a vector of length N, then what will be the length L of the vector, used to represent $y(t)$, that is returned by the **conv** function?

Question 4.   (a) Express $H(j\omega)$ in terms of the spectrum of any binary waveform $P(j\omega) = \mathcal{F}\{p(t)\}$. Then determine the matched filter output in terms of $P(j\omega)$ for a logical 0 and 1.

(b) For the matched filter impulse response you found in Question 1, determine the frequency response $H(j\omega)$.

(c) Again using the matched filter impulse response you found in Question 1, compute the spectrum of the matched filter output for both logical 1 and 0 inputs and sketch its magnitude and phase. Explain how the matched filter works in the frequency domain.

Question 5.   Write a MATLAB function file that will return the number of floating point operations (flops) used to compute the output of a matched filter for each of the implementation methods described below. You probably will want to use the MATLAB function **flops**. You will use this function in the laboratory problems to compare the relative computational complexity of these methods. Assume that the unit impulse response is represented by a vector of length N and the input signal by a vector of length M.

Write a MATLAB function that performs convolution (do not use the **conv** function), and write a MATLAB function to perform convolution using the DFT (implemented as **fft**) and the IDFT (implemented as **ifft**). You will use these two functions to compare the performance of the two methods. The **conv** function currently does not correctly report the number of floating-point operations required for the convolution.

Compute the number of floating-point operations for the following functions:

(a) Your straight convolution function

(b) Your **fft**-based convolution function using a length $L = M + N - 1$ DFT.

(c) Your **fft**-based convolution function using a length $L = 2^n = M + N - 1$ DFT.

8.4 BACKGROUND

Digital Communication Systems

In modern digital communication systems, information—including speech, video, and computer data—is encoded as sequences of binary ones and zeroes prior to transmission. Since continuous-time methods are necessary for transmission, the digital bits are encoded as continuous-time waveforms for transmission. For example, a 1 may be represented as a positive rectangular pulse, and a 0 as a negative rectangular pulse, each of duration T seconds.

In general, noise corrupts the signal during transmission. This noise may arise from many sources including, among others, interference from another transmitter and inclement weather conditions. The job of the receiver is to accurately discern what information was transmitted, even in the presence of this noise. The advantage of binary communication systems is that when the information is distorted by noise during transmission, the receiver needs only to determine whether a logical 1 or a 0 was sent, rather than to correctly identify all values of the original signal. The binary data can also be encoded so that any errors can be detected, accounted for, and possibly corrected.

While digital formats are very useful for transmitting information in the presence of noise, converting information such as speech, which is more naturally modeled by a continuous-time, analog signal, results in some degradation of the original signal. Specifically, the continuous-time signal is sampled, and each sample is digitally encoded using a number of bits that can fit in the allocated channel bandwidth. If the signal is bandlimited to less than half the sampling rate, then the signal can be exactly reconstructed from its samples; otherwise, controlled distortion in the form of a lowpass filter will be introduced prior to sampling. If the original signal has significant power at frequencies greater than half the sampling rate, then this process will introduce distortion. Furthermore, the representation of signals as binary data streams requires that the signal amplitudes be limited to distinct levels; this process is called quantization and introduces quantization noise.

So the question is: why use digital communication systems, knowing that distortion will be introduced? The answer is that the degradation resulting from sampling and quantization is controllable and can be designed to minimize its perceptual effects. Conversely, signal degradation that occurs during transmission is unknown and changes with time and thus cannot be controlled. With digital transmission, the digital bits are correctly extracted from the noisy received signal and then used to exactly reconstruct the transmitted digital signal. When the digital signal is then converted back to the form needed for display or playback, it is of known quality since any distortion was introduced and controlled by the transmitter.

This is the same principle used to justify digital audio compact discs; the process of converting the recorded audio signal to digital form for CD storage introduces some controlled distortion, but the payoff is that—unlike classic records—CDs are very robust to changes in temperature and rough treatment. Thus you can expect high quality sound from your CDs for many years. In a similar manner, digital transmission systems are more robust than their analog counterparts.

Noise Reduction Using Correlation Receivers

The problem of recovering digital signals corrupted by noise has many solutions. The correlation receiver can be shown to be the best solution for recovering digital data encoded using a known, but arbitrarily shaped, waveform in the presence of additive white Gaussian noise. A correlation receiver basically compares the received signal to the waveforms it expects to receive, each of which map to known bit representations. For example, a 1 may be represented by a 5-V signal for 1 second and a 0 by a –5-V signal for 1 second. The correlator tries to match intervals of the received signal to one of the expected binary waveforms, either 5-V or –5-V, and selects the best match. For this example, a positive number would indicate a 1 and a negative value would indicate a 0. Such a system is illustrated in Figure 8.4.1.

For example, a +5 signal, when multiplied by itself, yields $25(u(t) - u(t - 1))$; when integrated over 1 second, it gives a value of 25. Conversely, a –5 signal, when multiplied by +5, yields $-25(u(t) - u(t - 1))$; when integrated over 1 second, it gives a value of –25. The receiver chooses between +25 and –25 to determine 1 and 0.

The noise, $n(t)$, is modeled as a zero-mean, unit-variance Gaussian process that is uncorrelated with the signal—the classic AWGN (Additive White Gaussian Noise) model. This model is reasonable for many types of physical distortions, such as television static. The receiver works on the principle that since the signal is integrated over time, the continuous averaging process will eliminate the noise since it has a zero average.

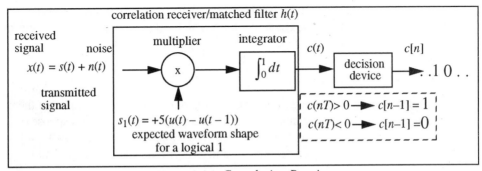

Figure 8.4.1 *Correlation Receiver*

The correlator receiver as shown is extremely difficult to build with a circuit, since an analog multiplier is quite complicated. An easier implementation of this receiver that gives the same result is the matched filter receiver. This receiver works on the principle that a linear system, which can be easily implemented, can be designed to perform the same operation as the correlator multiplier and integrator.

Implementing Correlation Receivers as Matched Filters

In the correlation receiver we essentially "slide" an expected pulse shape across the incoming signal and continuously multiply and then integrate to look for a match. Note that this operation sounds like convolution. However, in convolution, we flip the system impulse response in time prior to performing the multiplication and integration. It seems logical that if we simply define the impulse response of the filter in our receiver to be a flipped version of the expected binary waveform, then the process of convolution will flip it back. The resulting operation will exactly mimic that of the correlation receiver.

Mathematically, here's how it works. As you know, the output of a CT LTI system is given by the convolution integral:

$$c(t) = \int_{-\infty}^{\infty} x(\tau) h(t-\tau) d\tau$$

where $h(t)$ is the unit impulse response and $x(t)$ is the input. If we select $h(t)$ to be a flipped version of the expected waveform shape for a logical 1, shifted by the bit period T to ensure that the system is causal, then $h(t) = s_1(T-t)$ for t in $[0,T]$ and the convolution integral becomes

$$c(t) = \int_0^T x(\tau) s_1 (T-t+\tau) d\tau$$

Now, if we sample $c(t)$ at time T, the convolution integral becomes

$$c(t)|_{t=T} = \int_0^T x(\tau) s_1 (\tau) d\tau$$

which is the same mathematical operation as that performed by the correlator receiver! Thus, with a linear system, we can match a correlator receiver output by defining a causal CT LTI system to have an impulse response equal to a flipped and shifted version of the waveform representing a logical 1. Such a CT LTI system is called a matched filter.

Frequency Response of Matched Filters

Further understanding of matched filters can be gained by exploring the impact of the matched filter frequency response on the input signal spectrum. Since the impulse response of a causal matched filter can be expressed in terms of the expected binary waveform as $h(t) = s(T-t)$, then its frequency response $H(j\omega)$ can be expressed in terms of the waveform spectrum

$$S(j\omega) = |S(j\omega)|e^{j\omega \angle S(j\omega)}$$

Using this notation, the matched filter frequency response can be seen to correspond to the complex conjugate of the waveform spectrum

$$H(j\omega) = S(-j\omega) e^{-j\omega T} = |S(j\omega)|e^{-j\omega \angle S(j\omega)} e^{-j\omega T} = S^*(j\omega) e^{-j\omega T}$$

with a linear phase term as a result of the delay used to ensure a causal filter.

In this case, if we assume that the signal is just a single logical 1, i.e. $x(t) = s(t)$, then the matched filter output is

$$c(t) = h(t) * x(t) = s(t) * s(T-t)$$

for which the frequency spectrum is

$$C(j\omega) = |S(j\omega)|^2 e^{-j\omega T}$$

The sampling operation can be modeled in the time domain as multiplication by a periodic train of CT unit impulses, corresponding to convolution with a periodic train of impulses in the frequency domain:

$$c(t) \sum_{k=-\infty}^{\infty} \delta(t-kT) \Leftrightarrow |S(j\omega)|^2 e^{-j\omega T} \frac{2\pi}{T} * \sum_{k=-\infty}^{\infty} \delta\left(\omega - \frac{2\pi k}{T}\right)$$

Fast Matched Filter Implementation via the FFT

Implementing digital filters using convolution in the time domain is not computationally efficient for applications having long input signals or long unit impulse responses. Fortunately, convolution can be implemented as multiplication in the frequency domain, and there exists a computationally efficient method for digitally calculating Fourier transforms called the Fast Fourier Transform (FFT). The FFT implements the Discrete Fourier Transform (DFT), discussed in Laboratory Assignment 3, with the minimum number of multiplications, and is particularly efficient if the length used for the DFT computation is a power of 2. In MATLAB, the **fft** command will compute the DFT of any specified length as efficiently as possible.

Recall that convolving a sequence **x** of length M with a sequence **h** of length N results in a signal **y** of length $L = N + M - 1$. However, the DFT of a sequence has the same length as that sequence. Thus the length of the DFT used must be L. Performing convolution using multiplication of the sequence DFTs can be accomplished using the following steps:

1. Compute **X**, the length $P \geq L$ DFT of **x**.
2. Compute **H**, the length $P \geq L$ DFT of **h**.
3. Multiply to get **Y** = **HX**.
4. Compute the length P inverse DFT of **Y** to get **y**, the desired result.

If you select P to be a power of 2, then the convolution operation will require very low computational complexity. However, if you select P to be a prime number, then computing convolution in this manner will be quite computationally expensive.

8.5 LABORATORY EXPERIMENT

Sonar Detection

You are a sonar operator on a submarine and are told to find out where a friendly ship is with active sonar. Active sonar operates by sending a "ping" of sound and listening for return echoes from objects. Those of you who have seen *Hunt For Red October* may recall that active sonar is used in one of the scenes.

The sonar signal you send is a sinusoid of 0.25 second duration. Assuming sound travels 5000 feet per second in water, an object's distance can be calculated by measuring the return echo time. Unfortunately, a beach is near, and there is a lot of noise in the water.

The signal that was sent and the received signal are stored in the file **P_8_1.mat**. The transmitted signal is **x**, the received signal is **y**. A 0.01 second sampling rate was used to create these signals for use in MATLAB.

Problem 1.   Load this file, and determine how far away the friendly ship is by using a matched filter receiver on the echo. Remember that a matched filter gives a maximum value at the END of a signal, since it is sampled at time *T*. *Hint*: find the appropriate *h(t)* for the matched filter, and use the **conv** command to find the filter output.

Problem 2.   Find the frequency response of your matched filter and the frequency spectra of the received and filtered signals. Does this information help you in determining the return echo time? Consider the slope of the unwrapped phase of the received signal in explaining your answer. Remember that the noise in the system will affect your results.

Digital Message Reception

The friendly ship you detected in Problem 1 now wants to send you a binary message using sonar. The binary message scheme you use calls for 1 values to use a triangle wave 0.5 seconds long with a positive initial slope, and 0 values to use a triangle wave of the same length with a negative initial slope. Again, there is a great deal of noise in the water making detection and identification of the signals sent potentially difficult.

Problem 3.   Two vectors are stored in the file **P_8_4.mat**: a sample of the waveform shape used to denote a logical 1 signal and a sampled version of the received signal. The transmitted signal is **x**, the received signal is **y**. A 0.01 second sampling rate was used to create these signals for use in MATLAB. Since many bits were sent, look for more than one peak. Determine what binary message stream was sent. Then decode the message into human-readable characters using the ASCII code, stored in **ascii.code**. What message was sent?

> Given your results above and answers to laboratory preparation problems, explain the benefit of using matched filters in digital communications. Provide an intuitive explanation of why these work well in recovering bits of information from noisy signals. Consider what might happen if the message signal was not encoded in a binary code prior to sending, e.g., you simply sent a speech segment to be recovered.

Digital Communication Signal Generation

Problem 4.   Generate an ASCII message up to 10 characters long and encode it in a binary stream. Convert the binary stream to an analog signal using a triangle wave 0.01 seconds long for a logical 1. Use a sampling rate of 8 kHz. Play the message through a set of speakers and have another group record the output with a microphone. They should decode your message, and you should decode theirs.

> How much noise was present in this system, and how well did the message decode? What could be done to improve message reception?

Correlation Receiver Implementation

Problem 5.   The system described in Figure 8.4.1, "Correlation Receiver," on page 85, does not have to be implemented as a matched filter. Some systems actually perform the correlation operation and the integration. Design an M-file that performs the correlation operation for each bit in a message signal and returns a stream of 1's and 0's, and verify that the results you got for Problem 3 are correct. Recall that correlation involves multiplication by the original waveform and integration over a period.

> Are the two methods equivalent? Did you get the same results? What problems could arise using this method?

Computational Complexity of Different Convolution Implementations

Problem 6.   The objective here is to use the MATLAB functions that you created in Lab Prep Question 5 to compare the computational complexity of the different methods that can be used to perform the convolution needed to compute the matched filter output. Create two simple signal vectors to be convolved using the different methods. In one vector create a signal of all ones that has 10 elements. In a second vector, create a ramp that goes from 0 to 1 in increments of 0.1.

(a) Using the MATLAB functions you developed in Lab Prep Question 5, compute the number of floating point operations needed to convolve these two functions using each of the different methods. Which is most efficient?

(b) Create a new MATLAB function that includes your original function inside a loop. In each iteration, add n additional ones to the vector of all ones so that its length is increased by n. In an output function vector, store the length L of convolution result for each iteration. In separate output vectors, store the number of flops required for each method to compute the convolution in each iteration.

(c) Graph the number of flops required for each method as a function of L on a single plot. Which method is most efficient for shorter-length convolutions? For longer ones? At what point(s) do the methods have the same computational complexity?

Laboratory Assignment 9

Filter Design for Audio Tape Restoration

PURPOSE

The purpose of this assignment is to introduce you to standard analog and digital filter design techniques and implementations. Your assigned task is to restore an audio tape that has been corrupted by an annoying hiss to its original quality. In this assignment, you will design and implement Butterworth filters that can be used to eliminate this audio tape distortion. Butterworth filters are widely used, and designs for lowpass filters are easily modified, via simple frequency transformations, to create highpass, bandpass, and bandstop filters. Butterworth filters may be implemented using either CT active filters (using operational amplifiers) or digital filters. Implementation with active filters requires determining the filter transfer function and selecting circuit parameters to implement it. Implementation with digital filters requires converting the CT transfer function to a valid DT transfer function using the bilinear transform. You will be asked to compare and contrast the performance of analog and digital Butterworth filters designed to restore the audio tape.

9.1 OBJECTIVES

By the end of this lab, you should be able to:

1. Define filter specifications to remove an audio tape distortion.
2. Design lowpass, highpass, and bandpass Butterworth filters to given specifications.
3. Implement Butterworth filters using CT active circuits.
4. Design digital Butterworth filters from a CT design using the bilinear transform.
5. Evaluate trade-offs of different filter designs and implementations for a given application.

9.2 REFERENCES

Review Topics

1. Filters and frequency response
2. Transfer function, poles, and zeros

3. CT and DT system frequency response

4. CT and DT system stability

5. Op-amp circuit analysis

Exploratory Topics

1. Butterworth filter definition and LPF design

2. Frequency transformations of Butterworth filters

3. Butterworth filter implementation with active circuits

4. Digital filter design using the bilinear transform

Application Reference

1. *Digital Filter Design* (Parks and Burrus, 1987, John Wiley & Sons Inc.)

2. *Discrete-Time Signal Processing* (Oppenheim and Schafer, 1989, Prentice-Hall Inc.)

3. *The Art of Electronics* (Horowitz and Hill, 1989, Cambridge University Press)

9.3 LABORATORY PREPARATION

Problems

Question 1.   (a) Analyze this circuit and determine its frequency response as a function of the circuit components.

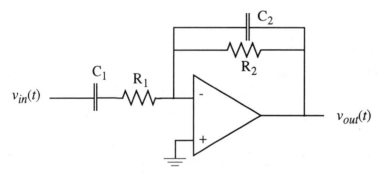

Figure 9.3.1 *Circuit for Question 1*

What relationship must the circuit components satisfy in order for the circuit to have a circuit gain of 2 and a passband ranging from 500 Hz to 2000 Hz? The general form for a second-order bandpass filter is

$$H(s) = \frac{Ks}{(s + \omega_l)(s + \omega_h)}$$

(b) Determine the frequency response in terms of the filter components if the capacitor C_1 is removed.

Question 2. Design a Butterworth filter that has a passband ripple of 3 dB and a stopband attenuation of 20 dB. The passband must reach to 900 Hz, and the stopband must start at 1.2 kHz. Determine how many sections and what cutoff frequency are required.

Question 3. Find component values for the circuit in Question 1(a) so that the passband ranges from 500 Hz to 2000 Hz, and the circuit gain is 2. Use $C_2 = 0.01$ μF, $R_1 = 3.2$ kΩ. Consider what components you can obtain, and what trade-offs must be made. Plot the magnitude of the frequency response over the range 20 Hz to 44 kHz to verify your solution.

Question 4. Determine the numerator and denominator coefficients for a digital filter implementation of the Butterworth filter you designed in Question 2. Find the frequency response of these coefficients using **bode**, and determine if your design met the specifications.

Question 5. Consider the fourth-order Butterworth filter shown in Figure 9.4.3.

(a) Use the graphical method for determining $|H(j\omega)|$ from its pole locations to verify that Butterworth filters are maximally flat.

(b) Determine the frequency response $H(j\omega)$ assuming a cutoff frequency of 100 Hz.

(c) Using ω_c from (b) and unity gain, determine the system transfer function $H_{LP}(s)$.

(d) Sketch or plot the magnitude of the frequency response for $H_{HP}(s) = 1 - H_{LP}(s)$. Verify that it is a highpass filter.

Question 6. Consider the circuits shown in Figure 9.4.4 and Figure 9.4.5.

(a) Determine the transfer function and frequency response for the circuit in Figure 9.4.4. Verify that it implements a lowpass filter by determining the magnitude of its frequency response for $\omega = 0$ and the limit as $\omega \to \infty$.

(b) Repeat part (a) for the circuit in Figure 9.4.5, and verify that it is a highpass filter.

9.4 BACKGROUND

In Laboratory Assignment 4, you explored the idea of ideal filtering. Using ideal filters, the frequencies in the passband appear in the output—all modified by the same gain and having the same time delay—while the frequencies in the stopband are completely eliminated; the bandwidth of the transition from passband to stopband is infinitely narrow.

In this assignment, you will learn how to design and implement both CT and DT LTI filters described by linear, constant-coefficient differential and difference equations, respectively. We will refer to these collectively as realizable filters. Ideal filters are non-causal and cannot be represented finite order differential or difference equations, and are therefore not realizable. From Fourier analysis, recall that creating a rectangular frequency function requires that the unit impulse response, either DT or CT, be an infinite sinc function. All future values of the input signal would be needed to determine the filter output at the current time.

Filters and Frequency Response

Filters are characterized by their frequency responses, which define how the filters modify the signal frequency content from input to output. Representative magnitude responses for realizable filters are illustrated in Figure 9.4.1. Note that the edge of the passband now is defined as the point where the magnitude of the system frequency response drops to $1/\sqrt{2}$ times its maximum gain of G. These points are called half-power or 3 dB points. At the filter output, the ratio of the power of a sinusoid at this frequency vs. that of a sinusoid at the frequency corresponding to maximum gain is $[(G/\sqrt{2})/G]^2=(1/2)$, or $10\log_{10}(1/2) = -3$ dB. Diferent ways to determine the system frequency response from different representations of LTI systems are discussed below.

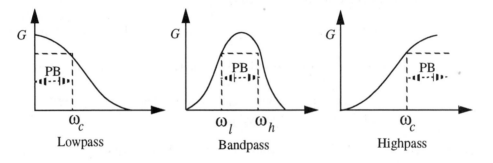

Figure 9.4.1 *Magnitude of Realizable Filters $|H(j\omega)|$ vs.* ω

CT LTI systems used as filters are characterized in the time domain by the system unit impulse response $h(t)$, and causal realizable filters can be characterized by the system transfer function $H(s)$ (see Lab 5). From these, the filter frequency response $H(\omega)$ is

$$H(\omega) = H(s)\big|_{s=j\omega} = \mathcal{L}\{h(t)\}\big|_{s=j\omega} = \mathcal{F}\{h(t)\}$$

which modifies the input signal spectrum $X(\omega)$, resulting in the output signal spectrum
$$Y(\omega) = H(\omega)X(\omega)$$

Such systems are asymptotically stable if all of the poles, or, equivalently, the characteristic roots, are in the left half-plane (where $\text{Re}\{s\} < 0$), and exhibit BIBO (Bounded Input, Bounded Output) stability. Any poles on the imaginary axis or in the right half-plane result in marginally stable or unstable systems, respectively; both of these cases correspond to BIBO unstable systems. If the input to a marginally stable system is a pure sinusoid at the resonating frequency, then the system output will become unbounded.

Conversely, DT LTI systems used as filters are characterized in the time domain by the system unit pulse response $h[n]$, and causal realizable filters can be characterized by the system transfer function $H(z)$ (see Lab 10). From these, the filter frequency response $H(\Omega)$ is:

$$H(\Omega) = H(z)\Big|_{z = e^{j\Omega}} = Z\{h[n]\}\Big|_{z = e^{j\Omega}} = \text{DTFT}\{h[n]\}$$

which modifies the input signal spectrum $X(\Omega)$, resulting in the output signal spectrum

$$Y(\Omega) = H(\Omega)X(\Omega)$$

DTFT stands for Discrete-Time Fourier Transform. You do not need to be familiar with the DTFT to complete this assignment; this mathematical relationship is included for completeness. In all assignments, your Fourier analysis will be performed using the DFT in MATLAB.

DT LTI systems are asymptotically stable if all of the poles, or equivalently the characteristic roots, are inside the unit circle, $|z| < 1$; such systems also exhibit BIBO (Bounded-Input, Bounded Output) stability. Any poles on or outside the unit circle result in marginally stable or unstable systems, respectively; both of these cases correspond to BIBO unstable systems.

For both DT and CT filters, the magnitude of the frequency response determines the gain applied at each frequency to obtain the output signal spectral magnitude, and the phase of the frequency response defines how much the phase of the input signal is shifted to obtain that of the output signal, i.e.,

$$|Y| = |H||X|$$
$$\angle Y = \angle H + \angle X$$

Essentially the phase of the frequency response defines the time delay that the filter will cause at different frequencies.

Goals and Trade-offs

Your goal is to design filters for removing a "hissing" distortion that is corrupting an audio tape. The CT signal from the audio tape has been sampled and digitized and is stored in a file. You may use either analog or digital filters to accomplish the assigned task; the choice between CT and DT filters depends upon your goals.

Using analog filters, you can process the audio signal directly from the audio tape without worrying about introducing additional distortion through sampling and quantization. If you expect to store the recovered signal on tape, then analog processing will allow you to record the restored signal onto tape as you process it. Analog filters implemented with op-amp circuits are inexpensive, low power, and easy to design and build. However, the filter performance will change over time due to slow degradation of the circuit components. If this is a one-time processing need, then analog filters present a cost-effective option. If you expect to have a need to process audio signals having the same, or only slightly different, distortion at some later date, then the time required to design and use digital filters may be more cost-effective.

While caution must be used to prevent the introduction of additional distortion caused by sampling and quantization, the ability to process your signal on a digital computer allows you to implement more mathematically sophisticated algorithms. In this assignment, you will be designing straightforward digital filters. However, there still are some advantages to digital processing. If you plan to store the recovered audio signal in digital format on a CD, then it makes sense to design a digital filter that can be used again if you encounter the same problem. By using a CD for storage, you will reduce the likelihood that the audio signal will be further distorted at some later date due to mishandling or poor storage conditions. Since you can implement the digital filter simply as a computer program, then you can exactly replicate its performance as needed.

Filter Specifications

Since you cannot implement ideal filters, it is important to have a means for specifying the performance required from your filter implementation. Once you know how performance is tied to characteristics of the filter frequency response, you can evaluate the performance trade-offs between different filter designs. Since the signal that you wish to process is a CT signal, we will express the filter specifications in terms of CT system frequency response. Later we will show how to design a DT transfer function that will meet the same specifications when used to process the sampled signal.

The terminology used to describe the frequency response characteristics that need to be specified actually makes a lot of sense. The standard notation and terminology for these characteristics are shown in the figure below for a lowpass filter:

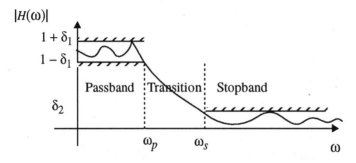

Figure 9.4.2 *Definitions for Filter Specifications*

The parameter δ_1 indicates how much ripple, or variation in gain, you will allow in the passband; ω_p is the frequency at the edge of the passband, defined as the point where the passband magnitude is decreasing and crosses through $1 - \delta_1$; δ_2 indicates how much attenuation is required in the stopband; and ω_s is the frequency at the edge of the stopband, defined to be the point such that gains at all frequencies beyond it in the stopband are attenuated to δ_2. Note that the gain in the passband, for the purposes of specifying δ_1 and δ_2, is normalized to one.

For highpass filters, these parameters are similarly defined. In the figure above, if you simply leave the vertical axis and labeling fixed, fix the horizontal axis (but not its labeling), and replace the remaining graphics by their mirror image, then you will have the definitions for a highpass filter. These parameters again are similarly defined for bandpass filters, but now there are two transition and stopband regions, so you must specify two sets of passband and stopband frequencies, one set for the upper transition band and one for the lower transition band.

In general, we would like filters to have very little ripple in the passband, have narrow transition bandwidth, and have large attenuation in the stopband, the same as ideal filters. Since achieving the ideal is impossible, a number of different techniques for designing filters have been developed that emphasize different sets of these objectives. By designing a filter, we mean determining a transfer function, or—equivalently—the system poles, zeros, and gain. One of the most common types of filter is called a Butterworth filter. Butterworth filters are maximally smooth, i.e., no oscillation appears in the magnitude of the frequency response in either the passband or stopband.

Butterworth Filter Design

The maximally smooth frequency response for a Butterworth lowpass filter is achieved by placing the poles in the semicircular arrangement shown below, where $2N$ is the number of poles.

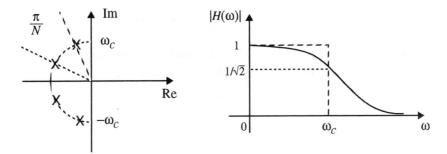

Figure 9.4.3 *Fourth Order Butterworth Filter Design Example*

Note that all of the poles have a magnitude of ω_c, and are equally spaced with an angular spacing between poles of π/N. Recall that the magnitude of the frequency response at a given frequency is inversely proportional to the product of the distances from all poles to that frequency on the imaginary axis. Using this intuition, you should be able to explain why this arrangement of poles provides such a smooth frequency response. Increasing the order of the filter makes the rolloff (slope) steeper, resulting in a narrower transition bandwidth.

Designing a Butterworth filter means selecting a filter order N and cutoff frequency ω_c such that the desired specifications are met. The magnitude of the frequency response for a Butterworth filter is given by

$$|H(j\omega)| = \frac{1}{\sqrt{1 + \left(\dfrac{\omega}{\omega_c}\right)^{2N}}}$$

and there are four steps to determining the filter order and the cutoff frequency given the frequency specifications as defined in Figure 9.4.2.

Assume that you are to design a filter given the desired passband ripple δ_1, stopband ripple δ_2, and cutoff frequency ω_c. Using the following steps, you can determine N and ω_c for a lowpass Butterworth filter to meet these specifications.

1. Since Butterworth filters are uniformly smooth, we know that the gain at the passband frequency ω_p is defined to be a value of $1 - \delta_1$. We can use this knowledge to generate the first design equation as follows:

$$|H(j\omega)|^2 = \frac{1}{1 + \left(\dfrac{j\omega_p}{j\omega_c}\right)^{2N}} = (1 - \delta_1)^2$$

$$\left(\frac{j\omega_p}{j\omega_c}\right)^{2N} = \frac{1}{(1 - \delta_1)^2} - 1 = K_1$$

2. Similarly, we know that the stopband attenuation δ_2 must be achieved at the frequency ω_s, resulting in the second design equation.

$$|H(j\omega)|^2 = \frac{1}{1 + \left(\dfrac{j\omega_s}{j\omega_c}\right)^{2N}} = \delta_2^2$$

$$\left(\frac{j\omega_s}{j\omega_c}\right)^{2N} = \frac{1}{\delta_2^2} - 1 = K_2$$

3. If we divide the first design equation by the second, we can solve for N in terms of K_1, K_2, ω_p and ω_s. This N must be rounded to the next highest integer.

$$N = \frac{\ln\left(\dfrac{K_1}{K_2}\right)}{2\ln\left(\dfrac{\omega_p}{\omega_s}\right)}$$

4. Given N, we can compute ω_c from the first design equation.

$$\omega_c = \frac{\omega_p}{K_1^{1/(2N)}}$$

By using this design method, we can determine how complicated the filter structure needs to be and which requirements should be relaxed to reduce complexity.

Once you have determined ω_c and N such that the design specifications are satisfied, then you can determine Butterworth filter pole locations p_i, as illustrated in Figure 9.4.3. The lowpass filter transfer function $H_{LP}(s)$ can be determined from these poles and the gain G desired for your application:

$$H_{LP}(s) \;=\; \frac{G}{\displaystyle\prod_{k=1}^{2N}(s-p_k)}$$

We can use this same design to determine the transfer function for a unity-gain Butterworth highpass filter $H_{HP}(s)$ having the same cutoff frequency and rolloff from

$$H_{HP}(s) \;=\; 1 - H_{LP}(s)$$

Active Filter Implementation

Analog filters are usually constructed from operational amplifiers, resistors, and capacitors. There are two main filter structures: the lowpass filter, shown in Figure 9.4.4, and the highpass filter, shown in Figure 9.4.5. These are inverting filters, which implies that the sign of the output is opposite that of the input.

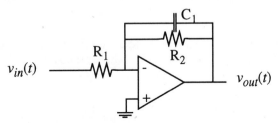

Figure 9.4.4 *Lowpass filter structure*

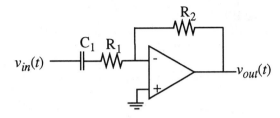

Figure 9.4.5 *Highpass filter structure*

The transfer functions for these filters can be found by using the generalized Ohm's law for passive circuit elements, $V(s) = I(s)Z(s)$; $Z(s)$ is the complex-valued impedance, and $V(s)$ and $I(s)$ are the Laplace transforms of, respectively, the voltage across and current through each passive circuit element. The complex impedances for resistors, capacitors, and inductors are, respectively, $Z_R(s) = R$, $Z_C(s) = 1/(sC)$, and $Z_L(s) = sL$. We will assume here that the op-amps are ideal, i.e., the voltage drop across and current into the input terminals are zero. The system transfer function can be found by performing nodal analysis at the negative input of the op-amp and solving for $H(s) = V_{out}(s)/V_{in}(s)$. Remember that you cannot apply nodal analysis at the op-amp output since the current out of the op-amp is unknown and cannot be assumed zero.

Digital Filter Design from Analog Filters

Discrete-time, digital filters also can be used to filter CT signals that have been sampled at a rate of f_s by simply writing a program which recursively solves the difference equation, given the initial conditions and input data sequence. Designing a digital filter then requires determining the coefficients of the difference equation, or, equivalently, the transfer function. How can we design a digital filter that satisfies the specifications expressed for the analog CT signal? Rather than going through an entirely different design procedure, we can simply convert the CT transfer function $H_{LP}(s)$ resulting from our design of a CT filter into a valid DT transfer function $H_{LP}(z)$ having equivalent performance. The process of converting a CT design into DT can be accomplished by the bilinear transform.

What do we need to consider when performing this transformation? Clearly, we need to consider the stability of the resulting filter. For stability, the poles of $H_{LP}(s)$ must be in the LHP, whereas those for $H_{LP}(z)$ need to be inside the unit circle. The sampling rate f_s also is important; it must be high enough to prevent aliasing and it limits the frequencies that we can represent to those that are less than $f_s/2$.

The bilinear transform maps the left half of the s-plane to the interior of the unit circle in the z-plane and is dependent upon $T_s = 1/f_s$. The bilinear transform is used to convert from a Butterworth CT filter design to an equivalent DT filter design as follows:

$$H(z) = H(s)\Big|_{s = \left(\frac{2}{T_s}\right)\left(\frac{1-z^{-1}}{1+z^{-1}}\right)}$$

Digital filters are categorized by the length of their unit pulse response: FIR filters have Finite Impulse Responses, and IIR filters have Infinite Impulse Responses. IIR filters may have both poles and zeros, while FIR filters have only zeros. In practice, FIR filters are widely used as they are simple to design and implement and are always stable. However, FIR filters typically must be of much higher order than IIR filters that provide equivalent performance, so FIR filter implementations generally have a higher computational cost.

Filter Analysis and Design Using MATLAB

In this assignment, you will use several MATLAB functions. The function **butter** designs a digital Butterworth filter to meet your specifications. Essentially, this function designs a CT

Butterworth filter and then uses the bilinear transform to obtain the DT filter design. The function **bilinear** will transform a CT filter design into an equivalent digital filter design using the BLT. The functions **freqz** and **bode** both are used to compute the frequency response for DT filters given the transfer function coefficients.

The **fir1** function tries to create a filter that matches samples of an ideal filter response meeting your specifications. We expect the resulting unit impulse response to have a $\sin(x)/x$ form, which has infinite length. The **fir1** function creates a filter by truncating this response to the number of samples requested. Using more samples improves the filter response but requires more computation. Both **butter** and **fir1** require filter order, a cutoff frequency, and a filter type (lowpass, highpass, or bandpass) as inputs, and return filter coefficients.

9.5 SOFTWARE EXPERIMENT

A sampled piece of music is stored in **P_9_1.wav**; its frequency spectrum is shown in Figure 9.5.1. When the same audio signal was recorded onto a tape, noise was introduced.

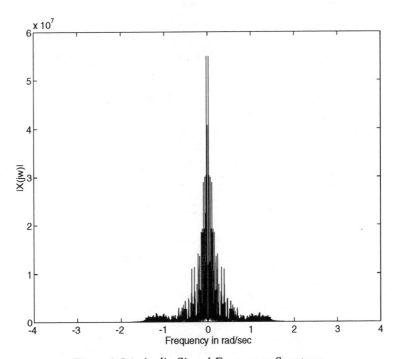

Figure 9.5.1 *Audio Signal Frequency Spectrum*

In the signal spectrum, this noise shows up as additional high-frequency signal energy. This sampled signal is stored in the sound file **P_9_2.wav**; you should play this sound file before beginning the problems below. The high-frequency hiss should be clearly audible.

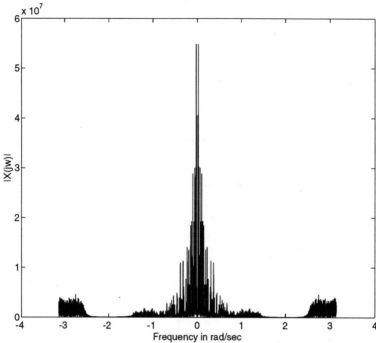

Figure 9.5.2 *Audio Signal Spectrum with Tape Hiss*

Problem 1.   Use the MATLAB function **bilinear** to transform your *H*(*s*) filter function from Question 3 into a digital filter. Verify, using **dbode**, that your digital implementation of the filter has the frequency response you expect.

> How well does the frequency response of the digital filter generated using this method match the frequency response you expected? Identify any differences.

Problem 2.   Filter the data contained in **P_9_2.wav** with the filter you created in Problem 1 using the **filter** command. Contrast the sound you hear with the noisy signal and the clean original signal. Look at the frequency spectrum of the filtered signal.

> How would you characterize the frequency content of the filtered signal? How, if at all, has it been modified by the filter? What would be an equivalent ideal filter and how would it affect the input signal spectrum?

Problem 3.   Examine the spectrum in Figure 9.5.2, and determine what cutoff frequency an ideal lowpass filter would need to remove the noise. Design a lowpass digital Butterworth filter that will remove the noise with 40 dB attenuation using the MATLAB function **butter**. Filter the data, and determine how well your filter did by comparing the original and filtered signals and their spectra.

> How does the digital Butterworth filter differ from the ideal lowpass filter? What trade-offs did you make to lower the number of coefficients? How much of the noise was removed? Does the filtering make a discernible difference in the audio quality?

Problem 4.   Design an FIR filter using the MATLAB function **fir1** that meets the same requirements as Problem 3. Use as long a filter as necessary. You will need to use **freqz** or **fft** to determine the frequency response of this filter and verify that it satisfies the required conditions.

> Which filter (FIR or IIR) does a better job filtering the signal? Which takes longer to perform the filtering? Which was easier to design, and which results in a response that is closer to the ideal? Contrast the frequency responses of your IIR and FIR filters to that of an ideal filter.

9.6 HARDWARE EXPERIMENT

Problem 5.   Construct the circuit shown in Figure 9.3.1 using the component values you found in Question 3, and add a variable gain stage as shown below in Figure 9.6.1. If necessary, add output transistors for larger current gain. Find the frequency response of your filter/amplifier system experimentally by using sinusoids of increasing frequencies as inputs, and then plotting the observed gain at the filter output as a function of frequency. Use frequencies in the range 20 Hz to 20 kHz. How would you characterize the behavior of this filter?

> What is the frequency response of the amplifier itself? Of the filter itself? Does the combination of both stages change the overall result? What happens to the output if a pair of headphones is attached to v_{out}?

Problem 6.   Remove the capacitor C_1 and see how the frequency response of the circuit has changed. Plot these new results and characterize the behavior of the filter. Contrast your results with those obtained in Problem 5.

Problem 7.   Play the file **P_9_1.wav**, which contains the data before it was put onto tape, through the filter. Comment on any audible changes in the sound introduced by the filter.

Problem 8.   Put the capacitor C_1 back into the circuit and comment on changes in the signal.

> What effect does the circuitry from Problems 7 and 8 have on the signal? How would you characterize the response of both systems? Is the effect noticeable? What would an ideal system do?

Problem 9.   Examine the spectrum in Figure 9.5.2, and determine what cutoff frequency an ideal lowpass filter would need to remove the noise. Design and implement a lowpass Butterworth filter that will remove as much of the noise as possible. Since a high-order Butterworth filter needs multiple stages, you will need to consider the hardware

---

requirements of your design. Comment on the changes your filter makes to the sounds in **P_9_2.wav**.

How different is the analog Butterworth filter from the ideal lowpass filter? What trade-offs did you make to lower the number of circuit elements? How much of the noise was removed? Does the filtering make a difference?

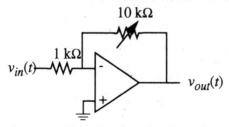

Figure 9.6.1 *Variable Gain Amplifier Stage*

Laboratory Assignment 10

Digital Filter Models for Acoustic Guitars

PURPOSE

In this laboratory assignment, you will design a digital filter to match the frequency response of an acoustic guitar body. You will explore how pole-zero locations impact the frequency response of digital filters and will use judicious placement of complex conjugate pole-zero pairs to replicate the desired frequency response. Using your filter, you will synthesize guitar sound by using appropriate excitations (inputs) that correspond to a plucked string. Recall that you can easily implement digital filters in MATLAB by using coefficients of the difference equation model and the **filter** command.

10.1 OBJECTIVES

By the end of this assignment, you should be able to:

1. Approximate the frequency response of a digital filter given the pole and zero locations in the z-plane.

2. Determine the impact of digital filters on the frequency content of sampled continuous-time input signals.

3. Design digital filters to synthesize the musical sounds generated by different instruments.

4. Predict how the output signal spectrum will look given the frequency response of the digital filter and the input signal spectrum, and how this effect impacts the signal audio properties.

10.2 REFERENCES

Review Topics

1. Frequency response of CT systems

2. Impact of poles and zeros on CT system frequency response

3. DT system characteristic roots and responses

Exploratory Topics

    1. Frequency response of DT systems

    2. Dependence of DT system frequency response on poles and zeros

    3. Numerical computation of system frequency response and signal spectra

    4. Inverse filters and stability

Application Reference

    1. Mathematical modeling of musical instruments: Fletcher and Rossing, *The Physics of Musical Instruments*. New York: Springer-Verlag, 1991

10.3 LABORATORY PREPARATION

Problems

Question 1.   A digital filter is described by the difference equation.

$$y[n+2] - \cos\left(\frac{\pi}{4}\right)y[n+1] + 0.25y[n] = 2x[n+1]$$

Determine the transfer function $H(z)$ and plot the poles and zeros for this DT system in the complex z-plane.

Question 2.   Suppose that the system from Question 1 is cascaded and has a transfer function $1/H(z)$ as shown.

In this case, express $y[n]$ in terms of $x[n]$. *Hint:* First express $Y(z)$ in terms of $X(z)$. What conditions must $H(z)$ satisfy for both of the systems $H(z)$ and $1/H(z)$ to be asymptotically stable?

Question 3.   A CT signal is sampled at 44.1 kHz. What CT frequencies (ω) correspond to the DT frequencies $\Omega = \pi/2$, $\Omega = \pi$, and $\Omega = 2\pi$. How does your answer change if the sampling frequency is 8 kHz?

Question 4.   For the second-order (2 poles) filter having pole/zero pair locations as shown in Figure 10.4.3a, determine how altering the arrangement of the pole/zero pairs impacts the magnitude of the frequency response. Assume that the poles always lie on the same radial line as the zeros, but allow their distances with respect to the unit circle and each other to vary. Note that either the pole or zero may be closer to the unit circle, and the zeros can lie outside the unit circle. How does the movement of these pole/zero pairs along this radial line and relative to each other impact the gain and bandwidth of the resonant peak or antiresonant notch? Sketch representative curves.

Question 5.   In Question 2, assume that $H(z)$ is a second-order system having two poles and two zeros, where the poles are located inside the unit circle at $z = 0.4 \pm j0.3$, and the zeros are located outside the unit circle at $z = 1 \pm j0.5$. In this case, the inverse filter $1/H(z)$ is unstable. Sketch the magnitude of the frequency response of the inverse filter. Now sketch the magnitude of the inverse filter when the poles are replaced by stable poles having the same angle as the original inverse filter poles but having magnitude equal to the reciprocal $1/|z|$ of their original magnitude. What do you notice?

10.4 BACKGROUND

Music Synthesis

Electronic music synthesizers are widely used for generating popular music and audio effects in theater productions. The majority of you have probably seen or heard a synthesizer at some point. Most synthesizers are based on one of two approaches: sampling or physical modeling. Synthesizers based on physical modeling attempt to mathematically describe the physical attributes (including nonlinearities) of actual instruments and, by exciting these models using appropriate input signals, generate the desired sound. Typically the accuracy of individual notes produced by these synthesizers is poor, and it is as difficult to produce high-quality sound on the synthesizer as it is to produce it using the actual instrument. However, sounds from notes played simultaneously or sequentially are blended together in much the same way as they are when played on the instrument itself.

Most performance-quality synthesizers operate by recording and digitizing samples of actual sounds from different instruments. These sounds are stored and then played back in the specified order (by using key presses) to generate the desired sequence of sounds. Pitches that are not stored are created using digital pitch shifting. However, such pitch shifting may cause undesirable changes in some fundamental audio properties of the original signal. Recall what happened in the lab when you used your MATLAB function **half** to shift your speech sample up one octave; in addition to sounding higher in pitch, your speech also sounded "munchkinized." By using this simple decimation procedure, you changed not only the pitch of the input signal, but also the characteristic frequency response of the vocal tract system; in essence, you made it shorter so that its resonant frequencies were higher. Thus, improving quality in sampled synthesizers requires using less pitch shifting, which necessitates storing a larger set of musical samples, which is costly. While the acoustic quality of the individual sounds generated using this approach is high, musical expression, or blending of notes, is low.

The accuracy vs. expression trade-off is represented by the curve shown in Figure 10.4.1.

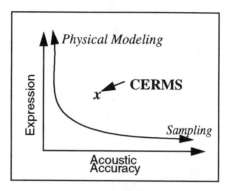

Figure 10.4.1 *Music Synthesis Curve*

You work for a company that has come up with a new product idea for producing low-cost, high-quality electronically synthesized musical sound, having both good expressivity and high acoustic quality. This new system is called CERMS, which stands for Code-Excited Recursive Music Synthesis, and the hoped-for performance improvements are illustrated by the x in Figure 10.4.1. Essentially this method corresponds to a physically based computational model which is inexpensive and easy to implement.

The basic idea is that by modeling the excitation (e.g., a plucked string) and instrument system (e.g., guitar body) separately, we can change the pitch without affecting the characteristic response of the guitar body. Thus pitch shifting can be done over a wider frequency range, resulting in lower storage requirements. If we use simple digital filters to model both the string and guitar body, then this approach can be cost-effectively implemented in either hardware or software. The pitch (note frequency) is modeled by a simple digital filter that has an exponentially decaying sinusoidal impulse response, and the response of the instrument "body" is modeled by a LTI digital filter, or difference equation. Any nonlinearities in sound production are incorporated into the input signal used to excite the model. Changes in pitch are accomplished by modifying the frequency of the pitch filter; the body digital filter serves to blend the sounds produced by the excitation.

To achieve the same selection of sounds as sampling technology, CERMS requires significantly less storage as fewer excitations are needed to generate the full range of sounds. Additionally, the body filter provides enhanced expressivity, and the pitch filter enables changing the pitch without the distortion caused by standard digital pitch shifting.

Your job is to explore the potential of this new approach by designing a digital filter to model the frequency response of an acoustic guitar body. Your group already has determined that the pitch needs to be generated by two DT LTI systems that filter the vertical and horizontal components of the plucked string excitation (the string vibration has components both in the vertical direction—perpendicular to the body, and the horizontal—parallel to the body), as shown in Figure 10.4.2. The notation z^{-T} is a delay operator of T samples, i.e., $z^{-T}x[n] = x[n - T]$. The two string models essentially make copies of one original period of the plucked string sound with a decay factor so that the overall response has an exponential decay. The original period of the plucked sound carries all the information about unmodeled dynamics (such as how the string was attacked during plucking). The pitch period is indicated by the number of samples, in general a noninteger amount, occurring in one pitch period.

You have been provided with the experimentally obtained DT unit impulse response for the guitar body. You are to try to match this response as closely as possible by choosing the coefficients of a difference equation model (having a DT transfer function $H(z)$), while trying to keep the order of the model (number of coefficients) low. Modeling the guitar body is the most important part of the synthesizer; a guitar's characteristic sound is governed by the guitar body, which provides resonance and amplification of the plucked string sound.

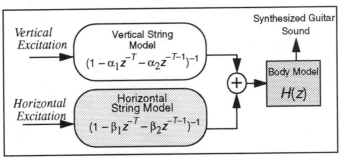

Figure 10.4.2 *Guitar Model*

Digital Filter Design and Frequency Response

Analogous to the CT transfer function $H(s)$, DT systems also are represented by a DT transfer function denoted by $H(z)$. The DT transfer function can be determined directly from the I/O difference equation describing a DT system, much in the same way that $H(s)$ is determined directly from a CT system I/O differential equation. Recall that for CT systems, we replaced the differential operator (d^n/dt^n) by s^n and solved for $H(s) = Y(s)/X(s)$. For DT systems, we replace each shifted input $x[n + N]$ or output $y[n + N]$ signal, shifted by N, by $z^N X(z)$ or $z^N Y(z)$, respectively. The DT transfer function $H(z)$ then is defined as $Y(z)/X(z)$. The functions $X(z)$ and $Y(z)$ are called the z-transforms of the DT signals $x[n]$ and $y[n]$ respectively, and are analogous to the Laplace transforms $X(s)$ and $Y(s)$ for CT signals.

Given a DT LTI system described by the Nth-order linear constant coefficient I/O difference equation

$$y[n + N] + a_{N-1}y[n + N - 1] + \cdots + a_1 y[n + 1] + a_0 y[n]$$
$$= b_0 x[n] + b_1 x[n + 1] + \cdots + b_M x[n + M]$$

the DT transfer function, assuming $y[n]$ is the output and $x[n]$ is the input, is

$$H(z) = \frac{b_0 + b_1 z + \cdots + b_M z^M}{a_0 + a_1 z + \cdots + a_{N-1} z^{N-1} + z^N} = \frac{Y(z)}{X(z)}$$

Again, analogous to the CT case, the value(s) of z that make the numerator of $H(z)$ zero are called DT system zeros, and the value(s) of z that make the denominator of $H(z)$ zero are called

DT system poles (characteristic roots). However, in contrast to the CT case, where negative real parts for poles guarantee system asymptotic stability, recall that the characteristic roots (poles) for the DT system must have magnitude less than one to ensure that the system is asymptotically stable.

For CT systems, the frequency response is determined by evaluating the system transfer function $H(s)$ along the marginal stability boundary $s = j\omega$, which corresponds to determining how the magnitude and phase of everlasting sinusoids (or steady-state nondecaying sinusoids) at different frequencies are altered by the system. Analogously for DT systems, the frequency response of the DT system is determined by evaluating the system transfer function $H(z)$ along the marginal stability boundary, i.e.,

$$H(z)\big|_{z = e^{j\Omega}} = H(e^{j\Omega}) \text{ for } (0 \le \Omega < 2\pi)$$

As in the CT case, the DT frequency response represents the response of the DT system to undamped DT sinusoids at different DT frequencies Ω. Recall that the frequency Ω of a DT sinusoid obtained by sampling a CT sinusoid at a sampling frequency of f_s is related to the frequency Ω of the original CT sinusoid by $\Omega = \omega T_s$, where $T_s = 1/f_s$. This result comes from the relationship between a CT sinusoid and its sampled counterpart $x[n]$:

$$x[n] = \cos(\omega t)\big|_{t = nT_s} = \cos(\omega n T_s) = \cos(\Omega n)$$

Given the pole and zero locations, the magnitude and phase of the DT frequency response at each DT frequency Ω can be predicted from the magnitude and phase of vectors connecting each pole and zero to the point on the unit circle corresponding to an angle of Ω. This procedure is analogous to the CT case where the magnitude and phase of the frequency response can be determined using the magnitude and phase of vectors from each pole and zero to each point on the imaginary axis, $j\omega$, representing the frequency response to a sinusoid having frequency ω.

Let zero locations be represented by the complex-valued numbers z_i, and poles by the complex-valued numbers p_i. Then

$$H(e^{j\Omega}) = \frac{(e^{j\Omega} - z_1) \dots (e^{j\Omega} - z_M)}{(e^{j\Omega} - p_1) \dots (e^{j\Omega} - p_N)}$$

If we define the difference vectors to be $d_{zi} = (e^{j\Omega} - z_i)$ and $d_{pi} = (e^{j\Omega} - p_i)$, then the magnitude and phase of the frequency response can be expressed in terms of the magnitudes and phases of these difference vectors:

$$|H(e^{j\Omega})| = \frac{\displaystyle\prod_{i=1}^{M}|d_{zi}|}{\displaystyle\prod_{j=1}^{N}|d_{pj}|} \qquad \angle H(e^{j\Omega}) = \sum_{i=1}^{M}\angle d_{zi} - \left(\sum_{k=1}^{N}\angle d_{pj}\right)$$

Designing digital filters to have the desired peaks and dips in the magnitude of the frequency response can be done with reasonable accuracy by placing complex-conjugate pole/zero pairs along the same radial lines, as illustrated in Figure 10.4.3a below.

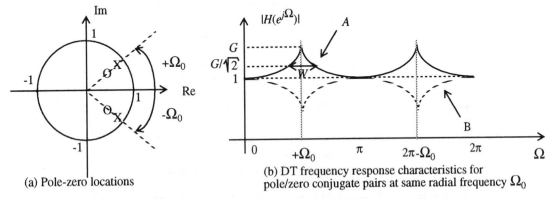

Figure 10.4.3 *Pole-Zero Placement to Control DT Frequency Response*

Two possible DT frequency response magnitudes are shown in Figure 10.4.3b. When the poles are closer to the unit circle than the zeros, the frequency response exhibits peaks at $\Omega = \Omega_0$ (curve A). Conversely, when the zeros are closer than the poles to the unit circle, the frequency response exhibits "dips" at $\Omega = \Omega_0$ (curve B). The gain G of the pole/zero pair is defined as the multiplier that yields the maximum magnitude of curve A (minimum magnitude for curve B). Often G is expressed in dB (decibels) as $20\log_{10}(G)$. The bandwidth W is defined as the "width" of the curve in radians where the magnitude is $1/\sqrt{2}$ times its maximum value, as illustrated in Figure 10.4.3b.

Note that the magnitude of the filter is close to unity for frequencies "far" from $\Omega = \Omega_0$. Multiple complex-conjugate pole/zero pairs can be placed as each pair only affects frequencies "near" the radial frequency of the pair. The frequency response for the total transfer function (having multiple pole/zero pairs) is then

$$H(e^{j\Omega}) = \prod_{i=1}^{N}\left(\frac{d_{zi}}{d_{pi}}\right)$$

Frequency Response of a Digital Filter in MATLAB

In MATLAB, digital filter difference equations of any order can be implemented using the **filter** command, as in **y = filter(b,a,x)**.

The frequency response of a digital filter can be found using the **freqz** command,

[h,w] = freqz(b,a,<# of points>, 'whole');

where **b** and **a** are vectors containing the filter coefficients; the number of points is the number of samples you wish to have on the frequency axis between 0 and half the sampling frequency; and "whole" indicates that we want to do both positive and negative frequencies. The vector **h** stores values of the complex frequency response for each frequency in **w**; as usual, we are interested in seeing plots of the magnitude and phase as a function of frequency. The function **freqz** determines the frequency response for all poles and zeros of the digital filter at equally spaced points along the curve $|z| = 1$ in the complex z-plane.

Implementation

Digital filters are usually implemented in software—a program is written that evaluates the difference equation describing the filter for every sample. MATLAB does just this when a call to the filter function is made. The coefficients of the difference equation are equivalent to the coefficients in the transfer function and can be found from the pole and zero locations by determining the numerator and denominator polynomials of the transfer function.

The only drawback to this method is numerical instability. Since a computer can only process numbers with finite precision, differences exist between the polynomial that has the desired roots and the polynomial representable by the computer. These errors can cause the filter poles defined by the implemented filter transfer function to be outside the unit circle, resulting in an unstable filter!

The solution to this problem is to use only polynomials of low order in implementing the filter, so that the numerical error is negligible. Second-order polynomials work quite well (each polynomial has two roots). Since the IIR filters are designed by pole/zero placement, it is not difficult to represent the whole filter as a cascaded series of second-order systems. In the equation below, each polynomial has two roots, and the N equations multiplied together have the roots of the desired transfer function.

$$H(z) = \prod_{i=1}^{N} \left(\frac{b_{i0} + b_{i1}z^{-1} + b_{i2}z^{-2}}{a_{i0} + a_{i1}z^{-1} + a_{i2}z^{-2}} \right)$$

As an experiment to verify this behavior, construct a polynomial from multiple roots, and then use **roots** to find the roots of this polynomial (e.g., **roots(poly([1:20]))**). Note the magnitude of the error between the true roots and the calculated roots.

Your job is to generate and test a digital filter that models the magnitude of the frequency response for the body of an acoustic guitar. Generating this filter requires choosing both the number and location of the poles and zeros that provides the best match to an experimentally obtained guitar body response. We are providing you with the experimentally obtained digitized impulse response for an acoustic guitar body and have developed some MATLAB code to facilitate your design. Ideally, you want to use the lowest-order (N) model possible that still is able to capture the perceptually important acoustic properties of the guitar body's impact on plucked string excitations.

For example, your auditory system is more sensitive to changes in magnitude than to changes in phase; hence we are asking you to try to match only the magnitude response. The major resonant "peaks" and antiresonant "dips" in the low frequency (0-500Hz) region will be more critical to the perceived quality of your synthesized sounds than the spectral shaping at higher frequencies. Thus you may want to use a radial pole/zero complex-conjugate pair to try to accurately match the gain and bandwidth of each of these low-frequency peaks and dips, while using fewer pole/zero pairs to model the higher-frequency response.

Once you have generated a guitar body filter to model the sampled impulse response of the guitar body by using pole/zero placement, you will need to test your design. You will want to compare the unit impulse response and frequency response of your model with that of the guitar body. Also, you will need to compare the perceived audio quality of the output of your model in response to sinusoidal excitations (without using the pitch filter) and string model excitations (obtained by using vertical and horizontal excitations as input to the pitch filters as shown in Figure 10.4.2) to an experimentally obtained digitized guitar sound. We have provided you with the digitized recording of a plucked guitar G string for testing and comparison purposes. The fundamental frequency of a guitar G string is 196 Hz.

Preliminaries

Data (.mat) Files:

> **impulse.mat**: contains the experimentally obtained impulse response of the guitar body, sampled at 44.1 kHz, and will be used by **manual.m** to aid you in designing your filter.

MATLAB Function (.m) Files:

> 1. **manual.m**: the manual pole/zero placement function that will help you design your filter. Using this function, you specify the gain and bandwidth you want at specified radian frequencies by using the mouse, and **manual.m** returns the pole/zero locations and transfer function resulting from your specifications. More details are provided below.

2. **generate.m**: a function that automatically generates a plucked string excitation and uses this excitation as input to your guitar body model to synthesize the guitar sound.

3. **invert.m**: a function that generates the inverse filter $1/H(z)$ given the coefficients of $H(z)$. This function checks pole locations of $1/H(z)$ and modifies them as needed to be sure that they are inside the unit circle.

4. **cascade.m**: a function that performs the filtering operation for a set of cascaded second-order systems.

5. **makedimpulse.m**: a function that constructs the impulse response of a set of cascaded second-order systems.

6. **frequency.m**: a function that finds the frequency response of a set of cascaded second-order systems, similar to **freqz**.

Sound (.wav) Files:

> **guitg.wav**: digitized recording of a plucked guitar G string, sampled at 44.1 kHz, to be used by you for comparison to your synthesized sound.

The function **manual** takes the impulse response and a current filter implementation (as a set of cascaded second-order systems) as inputs and provides you with the numerators and denominators for your designed filter as output. It works entirely in discrete time: frequency is in radians and ranges from 0 to 2π. Use **help** in MATLAB to get more information.

When you start **manual**, it will operate as follows, prompting you for any needed information:

1. It calculates the frequency response of the experimentally obtained impulse response and displays the magnitude.

2. It asks you for axis corrections (regions to display). Since the frequency content is ranging from 0 to 2π but the interesting stuff is generally in the range 0 to 1, correct the axis as necessary.

3. It asks if you want to reset the numerator and denominator. If yes, the numerator and denominator are set to 1. Otherwise the filter is not changed. This allows you to stop and restart without losing information, or to keep a variety of filter designs around.

4. The frequency spectrum of the impulse response and the current filter response are displayed. When prompted, click on the graphics screen where you would like a peak or dip in the frequency spectrum to occur. The function evaluates the frequency and gain of the point you entered and prompts you for acceptance.

Laboratory Assignment 10

5. The function calculates the change in dB from where you clicked to the current filter response, and it attempts to find the bandwidth of a peak or notch in the frequency response at that point.

6. The function then asks if you want to use the gain it calculated; if not, you can enter a desired gain (in dB).

7. The function then asks if you want to use the bandwidth it calculated; if not, you can enter a desired bandwidth (in radians/sec). This parameter controls roughly how tight the peak or dip generated by the filter is. Experiment some until you have an idea of what various bandwidths do.

8. The new frequency response is calculated and displayed. If the result is as desired, answer yes at the prompt.

9. If you want to add another point, answer appropriately and proceed to step 4.

10. When you exit, the pole and zero locations are plotted for you in polar coordinates in the complex z-plane.

Tasks

Problem 1.   Using **manual**, design a digital filter to match the frequency magnitude response of the guitar body impulse response. You will not be able to match it 100%, so come up with your best approximation using as few pole/zero pairs as possible. We suggest that you try to match only the general envelope of the frequency response, i.e. try to match only those features that would still be present if you "smoothed" the frequency response.

> What trade-offs did you make in approximating the frequency response? How did you determine if your response was "good enough"?

Problem 2.   Compare the impulse and frequency response of the guitar body and your body model. Use **makedimpulse** to get the response of your filter. To explore the effect of your filter, listen to the original **guitg.wav**. Use this signal to excite the *inverse* of your filter (generated using **invert**). Listen to the resulting signal; it is the input needed to generate the desired guitar sound using your body model.

> What can you discern about the success of your model by listening to this "residual" signal? How does this residual compare to the synthesized string signal?

Problem 3.   Create a sawtooth tone having pitch 100-300 Hz and listen to both this signal and the output resulting from using this signal to excite your body filter. Note similarities and differences. Does the filtered signal sound more guitar-like?

> What is the frequency content of the filtered sawtooth waveform? Compare and contrast your filtered waveform to the sampled waveform in both the time and frequency domain. How does your filter change the tonal quality of the sawtooth waveform?

Problem 4.   Using **generate**, create a guitar string excitation based on the model for a vibrating string. Use this signal as the input to the guitar body filter you designed and generate the resulting output. Listen to both the unfiltered string and synthesized guitar sounds, and compare these to the original sampled guitar sound. The string excitation alone can be created using **generate** by setting the numerator and denominator coefficients of the body filter to **[]** (null values).

> Attempt to quantify the resulting output: how much does it sound like a real guitar string? How much does it resemble (in frequency content) the sampled guitar sound? What changes still need to occur? If you convolved the original sampled impulse response with the string sound, would the results be different? Better? Slower?

Problem 5.   Try synthesizing a tune using pitch shifting. Recall how various notes relate to one another from Table 2 on page 27. The MATLAB functions **decimate** and **interpolate** will be useful for shifting pitch by less than an octave. To shift pitch by noninteger multiples, interpolate to the Least Common Multiple and then decimate to the desired result. For example, to change from 200 to 300 Hz (up by a factor of 1.5) interpolate by 2 and then decimate by 3.

Laboratory Assignment 11
The Ball on a Beam Control Problem

PURPOSE

This lab introduces the motivation for and implementation of control systems in the context of automatically controlling the location of a steel ball on a pair of metal rods called the beam. Starting from basic time- and Laplace-domain system concepts, you will be introduced to standard PID (proportional-integral-differential) controllers and explore how effective these can be in controlling the location of the ball on the beam. You will explore how a controller, designed under the assumption that the ball and beam system is linear, needs to be modified to achieve the desired performance given the nonlinearities in the physical system. The impact of the system transfer function, use of feedback, and importance of the pole placement on system responses are discussed and explored within this context.

11.1 OBJECTIVES ——■——

By the end of this laboratory assignment, you should be able to:

1. Determine the overall transfer function of a control system, including both the controller and the system to be controlled.

2. Generate and plot poles and zeros of the transfer function.

3. Use placement of characteristic roots to generate the desired system response and ensure stability.

4. Interpret root locus diagrams and use them in designing control systems.

5. Identify unmodeled dynamics (i.e., nonlinearities) in the plant and compensate for these in your controller design.

11.2 REFERENCES ——■——

Review Topics

1. Transfer functions of CT feedback systems

2. CT response of second-order systems

3. Poles, zeros, and characteristic roots

---

1. Root locus

2. Feedback control concepts

3. PID controllers and design

4. Plant nonlinearities and impact on design

11.3 INTRODUCTION

—■—

Electrical engineers often find that they need to use circuitry or computer systems to control a physical system. Examples include air conditioning systems, where a thermostat turns a system on and off in response to changes in measured ambient temperature; robots, where sensors on the manipulator aid in positioning, grasping, and moving objects; and automobile cruise control, which maintains a set speed over changing road conditions. Such technologies are called feedback control systems; these systems use feedback from sensors in determining what action should be taken to achieve some desired goal.

In this assignment, we will provide you with a graphical simulation package so that you may investigate how changing parameters in a specified control system affects the location of a ball on a beam as you attempt to move it towards some desired position within a specified time.

11.4 LABORATORY PREPARATION

—■—

Problems

Question 1.   Verify that the transfer functions given for the systems shown in Figure 11.5.1a and b are correct. In each case, what are the physical implications of $k = 0$? In other words, what is the system zero-state response $Y(s)$ for a given input $X(s)$ when $k = 0$?

Question 2.   Write the state-space equations for the ball and beam system in standard matrix/vector form. Identify explicitly A, B, C, and D. From the state-space formulation, generate the transfer function $C(sI - A)^{-1}B + D$ and verify that it is the same as the transfer function obtained through block diagram reduction of the system shown in Figure 11.5.3. *Hint*: Use **ss2tf**.

Question 3.   Determine the overall transfer function for the ball and beam control system shown in Figure 11.5.4.

Feedback and Control

A control system consists of a controller and a plant (the system to be controlled) connected in a feedback configuration. The two standard feedback configurations are shown below:

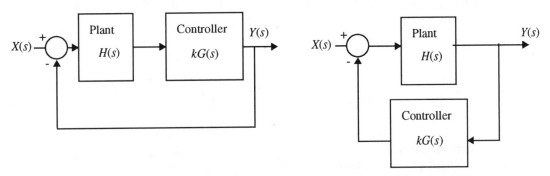

Figure 11.5.1 *Two Controller Configurations*

With some simple analysis, you should be able to determine that the transfer functions for the systems in Figure 11.5.1 are, respectively,

1.
$$\frac{Y(s)}{X(s)} = \frac{kG(s)H(s)}{1 + kG(s)H(s)} \qquad \frac{Y(s)}{X(s)} = \frac{H(s)}{1 + kG(s)H(s)}$$

The quantity $G(s)H(s)$ is called the open-loop transfer function, and $(1 + kG(s)H(s) = 0)$ is the characteristic equation for the closed-loop, or feedback, system. Solutions to the characteristic equation for different values of k yield the corresponding characteristic roots. Curves in the complex plane that indicate how these roots change as k is increased from zero are called root locus diagrams.

Using sensors to measure system output and then using this information to determine what action should be performed by the controller is called feedback. Often controllers are used as in Figure 11.5.1, where the goal is to match the system output $Y(s)$ to have some desired form $X(s)$. The feedback system is guaranteed to be asymptotically stable if all of the poles and zeros for the open loop transfer function are in the left half-plane. Consider analyzing the transfer function in Figure 11.5.1a for stability. If we take $G(s)$ and $H(s)$ to be represented by their numerator and denominator polynomial functions, we can express the system transfer function as follows:

2.

$$\frac{Y(s)}{X(s)} = \frac{k\dfrac{N_g(s)\,N_h(s)}{D_g(s)\,D_h(s)}}{1 + k\dfrac{N_g(s)\,N_h(s)}{D_g(s)\,D_h(s)}}$$

After multiplying through by a common denominator and simplifying, Equation 2 becomes:

3.

$$\frac{Y(s)}{X(s)} = \frac{kN_g(s)\,N_h(s)}{D_g(s)\,D_h(s) + kN_g(s)\,N_h(s)}$$

Note that the zeroes of the closed-loop feedback system are the zeroes of the open-loop transfer function (comprising the plant and controller zeroes), whereas the poles of the closed-loop system are variable and depend on the gain k. When k is 0, the closed-loop system poles are equal to the poles of the plant and of the controller. When k is large, the poles of the closed-loop system approach the zeroes of the open-loop transfer function; this can be seen by dividing both the numerator and denominator of $Y(s)/X(s)$ by k, letting k go to infinity and determining the poles of the resulting transfer function.

The Ball and Beam Problem

A metal ball is placed on a beam made of two parallel rods placed close together with ends attached, which is connected to a motor at its center. The motor is used to control the rotation of the beam about the beam's center point. The goal of the control task is to move the ball from the point where it is placed on the beam to a specified position. This task is achieved by changing the angle of the beam with respect to the horizontal so that the ball will roll downhill towards the desired position. The problem is that the ball has very little friction and has a tendency to keep rolling; as the ball approaches the desired beam position, the controller must make the ball stop by changing the angle of the beam to cause the ball to go uphill.

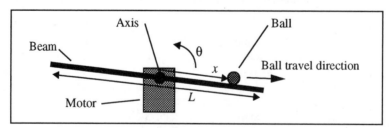

Figure 11.5.2 *Ball and Beam System*

Ball and Beam Physics

Designing a controller that will generate the correct signals for driving the motor to correctly position the ball requires understanding the physics and modeling of the ball and beam system. Consider the moment of inertia, J, of the whole system, which is the sum of the moments of inertia of all the moving parts of the system. For the ball and beam system, J is the sum of the moments of inertia of the beam and the ball. Define M_{be} to be the mass of the beam, M_{ba} to be the mass of the ball, L to be the length of the beam, and x to be the distance of the ball from the center of the beam as shown in Figure 11.5.2. Using this notation, the moment of inertia of the whole system is

4. Moment of Inertia
$$J = \frac{M_{be}L^2}{12} + M_{ba}x^2$$

The beam changes position due to the torque arising from two sources, the motor and the position of the ball coupled with the effect of gravity. We note that the overall torque (Γ) is equal to the moment of inertia of the system J times the angular acceleration of the system (α). Therefore, the torque is

5. Torque
$$\Gamma = \Gamma_m + M_{ba}x^2 g\cos\theta = J\alpha$$

where Γ_m is the torque generated by the motor, θ is the angle of the beam with respect to horizontal, and g is the gravitational acceleration, 9.8 m/sec^2. Angular acceleration is the torque divided by the moment of inertia J. Note that the motor torque can change sign depending upon which way the motor seeks to turn the beam.

The motion of the ball on the beam is determined primarily by its acceleration, a_{ba}, which is determined by the effect of gravity on the ball for a given beam angle:

6.
$$a_{ba} = -\frac{5}{7}g\sin\theta$$

From Equation 5 and Equation 6, we can represent the ball and beam system mathematically using a state-space representation with four state variables: ball position x, ball velocity v, beam angle θ, and angular velocity of the beam ω. The position of the ball is measured with respect to the center of the beam. The top two equations describe the motion of the ball along the beam, and the bottom two describe the angular motion of the beam.

$$\frac{dx}{dt} = v \qquad\qquad \frac{dv}{dt} = a_{ba} \cong -\frac{5}{7}g\theta$$

$$\frac{d\theta}{dt} = \omega \qquad\qquad \frac{d\omega}{dt} = \frac{1}{J}(\Gamma - M_{ba}gx)$$

We assume that the beam does not sway at large angles, so we can approximate $\sin(\theta)$ with θ.

The system transfer function can be found by solving the state-space equations above in the Laplace transform domain.

System Description as a Block Diagram

The block diagram model for the ball and beam system, as defined by the state-space equations above is illustrated below, where $X(s)$ is the output and $\Gamma(s)$ is the input.

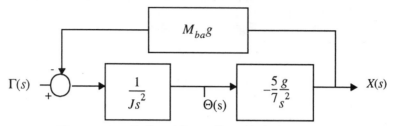

Figure 11.5.3 *Ball and Beam System Block Diagram*

Clearly the torque Γ feeds into the system and produces a beam angle θ (after integrating twice as indicated by $1/s^2$) and a ball position x, after integrating twice again. The position x also affects the torque as indicated by the feedback loop. As an example of the system behavior, consider the case where the input torque is 0; the ball position causes the beam to rotate so that the ball falls off the beam.

Control System Design

In designing a controller to efficiently move the ball to a desired beam position, we need to consider the initial ball position and velocity and the angle needed to move the ball to the desired position.

A standard type of analog controller is a PID controller. PID stands for proportional, integral, differential; i.e., gains (k_ρ), integrators (k_i/s), and differentiators (k_ds) are used to achieve the control objective: in this case, driving the motor so that it moves the ball to the specified beam position from its current location. By using a PID controller, we can continuously determine the control signal needed for the motor to generate the beam angle that simultaneously minimizes both the error between the ball's current x and desired x_d positions and the ball's velocity (so that it will come to rest at the desired location).

7.
$$\Theta_d(s) \ = \ \frac{7}{5g}\left(k_{pp} + \frac{k_{ii}}{s} + k_{vv}s\right)[X(s) - X_d(s)]$$

This PID controller equation describes the desired angle θ_d as being proportional to the difference in the position $x - x_d$, the velocity dx/dt, and the past history of the position errors, represented by the integrator ($1/s$). The value x_d is used as an input. Since this is a linear equation, it can be constructed using standard circuit elements.

How does this controller work to minimize the position error and velocity? If either the velocity or the position difference is large, the angle is made large to correct the errors. The integral term induces zero steady-state error. To implement the angle placement, let us use a similar PID controller, which will produce the torque necessary to move the beam to the desired angle.

8.
$$\Gamma(s) \ = \ M_{ba}gX(s) - J\left(k_p + \frac{k_i}{s} + k_vs\right)[\Theta(s) - \Theta_d(s)]$$

This controller generates the amount of torque needed to both move the beam to the desired angle and to overcome the moment of inertia due to the ball. Adding these controllers to the ball and beam system results in the overall system block diagram shown below:

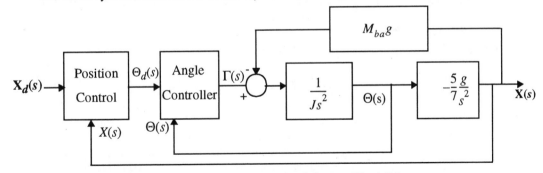

Figure 11.5.4 *Ball and Beam Control System Block Diagram*

Designing and implementing a controller for this ball and beam system is not an easy task. In this laboratory, you have the chance to explore controller design for this ball and beam system by selecting the gains of the PID controllers so that the desired task is performed. You then will test the performance of the controllers you design using a graphical simulation of the ball and beam system, called **beamsim**.

11.6 LABORATORY EXPERIMENT

Your task is to explore how the gains k_p, k_v, k_i, k_{pp}, k_{ii}, and k_{vv} impact system operation and stability. These are passed as arguments to the program **beamsim**. Additional arguments contain information about the physical system, including beam length, beam mass, ball mass, and desired ball position. Be sure to get the arguments in the correct order; if you are unsure, type '**beamsim -usage**' to display the correct order for the arguments. **beamsim** exists for PC and X Window systems; the X Windows version has a user interface that allows you to change parameters without having to exit the program.

```
beamsim <beam length in m> <beam mass in kg> <ball mass in kg>
<desired position in m> <kp> <kv> <ki> <kpp> <kvv> <kii>
```

The desired position is measured in meters with respect to the middle of the beam; to move the ball left of the middle, use negative values. All numbers are signed and may have as many decimal places as necessary.

The program will open two windows for you; the first contains a simulation of the beam, along with information about the current position; the other is a plot of the current and desired ball positions as functions of time.

System Analysis

A SIMULINK model of the system has been constructed and is in the file **P_11_1.m**. Use **linsim** to test your choice of gains and determine the system step response. To run the simulation, you will need to set variables **Mb** (the ball mass), **J** (without the dependence on ball position), **kp, ki, kv, kpp, kii**, and **kvv**, and create an input vector **[t xd]**, where **t** and **xd** are column vectors containing time and desired position at that time.

Problem 1.   Find the s-domain representation of the controllers and find the transfer function between x_d and x. Use MATLAB to find the poles of the complete plant and controller system (Figure 11.5.4) and determine how the various gains affect the stability of the system. Minimize the overshoot of the ball—how far over the desired position it travels—and the system settling time—how long the ball takes to "settle" at the specified position. Since this is a high order system, obtaining an underdamped response is somewhat difficult. Use the physical system parameters as defined in Problem 2.

The length of time to move the ball to within 0.5% of the desired position should be under 10 seconds. As a starting point, use the set of gains $k_v = 7$, $k_p = 26$, $k_i = 0$, $k_{ii} = 0.3$, $k_{vv} = 3.5$, $k_{pp} = 1$. Determine a gain k_{pp} that produces an underdamped response. You may wish to generate a root locus plot to help you select values of k_{pp} that result in different system performance. Since k_{pp} is not in the standard **rlocus** feedback configuration, you need to create your root locus by computing the system poles for different values of k_{pp}.

> Why would minimizing overshoot and system settling time be important? Consider the response using the initial set of gains and determine what could be improved. Also, think about the physical meaning behind a response: if it oscillates, why does it oscillate? Why does the ball not stop instantly? What does it mean to increase k_p or k_{pp}? k_v or k_{vv}? k_i or k_{ii}?

System Simulation

You will get different results from your analysis using the **beamsim** simulator instead of the SIMULINK simulator, which is based on the linear model, since **beamsim** includes some of the nonlinear effects. You will observe that the dynamics of the ball and beam system are somewhat different than that predicted by the linear system model. Try to place the ball in the following positions with the specified configurations:

Problem 2.   Length = 1 m, beam mass = 0.24 kg, ball mass = 0.05 kg; move the ball to –0.25 m.
Problem 3.   Same as (1), except move the ball to 0.4 m.
Problem 4.   Same as (1), except move the ball to –0.49 m

> What changed between SIMULINK and the **beamsim** simulations? How does this change the system response for your gains? What "tweaking" did you do? The gains k_p and k_{pp} are proportional gains. What is the controller trying to do using the error and these gains? What is it trying to do using k_v and k_{vv}? k_i and k_{ii}? What has the most effect on the system response, and how does that relate to the physical system?

Laboratory Assignment 12
Amplitude Modulation

PURPOSE

In this assignment, you will explore the use of digital computers for the analysis, design, synthesis, and simulation of an amplitude modulation (AM) system. In AM systems, an information signal is multiplied by a sinusoid at a much higher frequency. This technique makes it possible to transmit the information signal over the air via electromagnetic waves; in essence, using a higher-frequency signal to "carry" the information signal. You should be familiar with AM radio, which uses this technique; the frequency indicated on your AM radio dial for each station is the carrier frequency.

In this experiment, you will explore how modulation, particularly AM, impacts the signal spectrum and how this information is used to simulate the modulation and demodulation of an audio signal. In the process, you will learn about distortion resulting from undersampling, called aliasing, and how to prevent it. This will be particularly important in simulating an AM system as modulated signals have substantially higher-frequency components than audio signals. From this assignment, you should gain an appreciation for some of the practical trade-offs in designing AM system technologies and in using sampled CT signals for AM system simulation.

12.1 OBJECTIVES

By the end of this laboratory assignment, you should be able to:

1. Compute, display, and interpret frequency spectra of an amplitude modulated signal.

2. Design and use digital filters to prevent aliasing in your computer simulation of an AM system.

3. Use your understanding of frequency-domain analysis and digital filters to design and build a system for modulating and demodulating audio signals.

4. Prevent aliasing by using filters or changing the sampling rate.

12.2 REFERENCES

Review Topics

1. Frequency response and filtering

OBJECTIVES

2. CT Fourier analysis

3. Common Fourier Transforms and properties

Exploratory Topics

1. Amplitude modulation and demodulation

2. Sampling and aliasing

3. Digital Fourier analysis

12.3 LABORATORY PREPARATION

Problems

Question 1.   Your goal is to shift a 100-Hz tone up to 150 Hz using modulation. Determine the carrier frequency needed to accomplish this goal. Specify the filter, if any, needed to removed any unwanted tonal artifacts resulting from the modulation process. Assume that your filter is ideal, and specify its type and cutoff frequencies. Illustrate your understanding and verify your solution by sketching the spectra of the original signal, the modulated signal, and the modulated signal after filtering.

Question 2.   Assume that $x(t)$ is a signal having the frequency spectrum shown in Figure 12.5.1 with $B = 2\pi(100)$ radians per second. What is the minimum sampling rate ω_0 that can be used to sample this signal if no aliasing is to occur? Assume that this signal is now sampled at $\omega_s = 2\omega_0$. Sketch the magnitude of the frequency spectrum for the sampled signal. Specify the type, gain, and cutoff frequency of an ideal filter that can be used to recover an exact replica of the original signal from its samples.

Question 3.   Now assume that the signal in Question 2 is modulated using a 1 kHz carrier, resulting in the signal $x_m(t)$. What is the minimum sampling rate at which $x_m(t)$ can be sampled to ensure that no aliasing occurs? Sketch the spectra of both $x_m(t)$ and its sampled counterpart if the sampling rate is 2kHz. Will there be any aliasing distortion? Explain your answer.

Question 4.   Now assume that signal $x(t)$, as defined in Question 2, is amplitude modulated using a 1 kHz carrier, creating the modulated signal $x_c(t)$. Select reasonable values for the modulation index and carrier amplitude. The AM demodulator at the receiver is shown below:

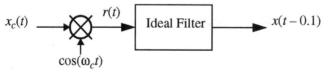

Figure 12.3.1 *AM Demodulator For Question 4*

Sketch $R(\omega)$. Specify an ideal filter such that the output is $x(t - 0.1)$, as shown, i.e., the output is an exact replica of the original transmitted signal shifted by 100 ms. Specify the filter type, cutoff frequency, and gain. Sketch the magnitude and phase of the filter frequency response.

12.4 INTRODUCTION

The purpose of this assignment is to further expose you to applications of digital Fourier analysis using MATLAB. These topics will be explored in the context of sampling audio signals for the design and simulation of the modulation and demodulation processes used in amplitude modulation. Since MATLAB is inherently discrete-time, you will explore practical considerations in the use of digital simulation and Fourier analysis for designing audio systems.

12.5 BACKGROUND

In the following sections, we discuss the effect of amplitude modulation and sampling on the signal spectrum using common properties of the Fourier transform. Both topics can be analyzed by understanding the effect of multiplying a signal by a periodic waveform. We use this information to provide an intuitive view of digital Fourier analysis as "sampling" of both the CT signal and its spectrum, and discuss digital filtering, frequency shifting, and aliasing as needed in your AM system analysis and design.

Frequency Shifting

How can we shift the center frequency of a sampled signal? One way is to use decimation and interpolation, as you learned in Chapter 3. Another way is to multiply your signal by a sinusoid in the time domain; this process is called modulation.

Recall that the Fourier transform of a sinusoid is a pair of unit impulse functions at positive and negative fundamental frequencies and multiplication in the time domain corresponds to convolution in the frequency domain. Thus, multiplication of a signal by a sinusoid results in the convolution of shifted δ functions with the input signal spectrum. The spectrum of the modulated signal will comprise our original signal spectrum shifted and centered around the positive and negative fundamental frequencies of the modulating sinusoid.

We can also understand the impact of this modulation processing in the time domain by using trigonometric identities. Suppose that your input signal is $\cos(\omega_1 t)$. If we want to shift this sinusoid to the frequency ω_2, we need to multiply by the frequency $\omega_2 - \omega_1$. From trigonometry we know that:

$$2\cos(\omega_a t) \cos(\omega_b t) = \cos((\omega_a + \omega_b) t) + \cos((\omega_a - \omega_b) t)$$

Our multiplication results in the signal $\cos(\omega_2 t) + \cos[(\omega_2 - 2\omega_1) t]$, while we wanted just $\cos(\omega_2 t)$. The additional unwanted sinusoidal component at $\omega_2 - 2\omega_1$ can be removed by high-pass filtering the modulated signal.

Aliasing

The term aliasing refers to the distortion that occurs when a continuous-time signal has frequency components at frequencies larger than half of the sampling rate. The process of aliasing describes the phenomenon in which power originally at these high frequencies appears as power in the sampled signal spectrum at a lower frequency, a frequency that can be represented using the given sampling rate.

One common example of aliasing is when wheels on a moving vehicle in a video appear to remain motionless or move "backwards," rotating in a direction opposite to that expected given the direction of travel. This effect can be understood using the idea of sampling: video is generated by displaying 30 still images, or frames, per second; this rate of display is high enough to trick the vision system into perceiving motion. Consider a wheel that makes 30 full rotations per second; this wheel will appear to be identical in each video frame. Thus the wheel appears to be motionless! Rather than seeing the high frequency rotation (30 rotations per second) of the wheel, it appears to be still, at a frequency of zero rotations per second.

Consider a CT signal for which the frequency spectrum in shown Figure 12.5.1. Note that the signal is bandlimited to $B/2\pi$ Hz.

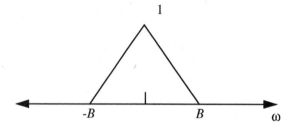

Figure 12.5.1 *Example Frequency Spectrum*

Sampling a signal can be modeled as multiplication by a train of periodic impulses in the time domain, corresponding to convolution of the signal spectrum with impulses at the harmonic frequencies of the time domain impulse train:

$$x_s(t) = x(t) \sum_{k=-\infty}^{\infty} \delta(t - nT_s) \Leftrightarrow X(\omega) \frac{1}{T_s} \sum_{k=-\infty}^{\infty} \delta\left(\omega - \frac{2\pi k}{T_s}\right)$$

If we sample this signal so that the sampling rate ω_s is less than $2B$—the Nyquist rate—then the frequency spectrum shown in Figure 12.5.2 results.

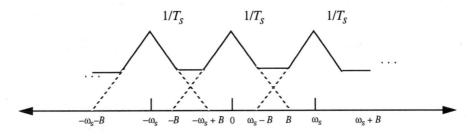

Figure 12.5.2 *Aliased Spectrum*

The overlapping sections add together, producing a distorted spectrum. Note the shape of the baseband (i.e. centered at zero) spectrum of bandwidth B; it is no longer perfectly triangular in shape, but is now "flat" in the band $\omega_s - B$ to B. The original signal spectrum cannot be recovered by using a lowpass filter, causing nonrecoverable distortion in the sampled signal.

Two methods are commonly used to prevent aliasing:

1. Sampling at a sufficiently high rate, that is $\omega_s 2B$.

2. Lowpass filtering the signal prior to sampling so that the signal to be sampled is band-limited to $\omega_s/2$.

The file **E_12_1.m** contains an example illustrating the effects of aliasing in the time and frequency domains.

Amplitude Modulation

Modulation is important because it allows us to transmit signals at different frequencies. Consider amplitude modulation radio systems: all audio signals are baseband signals comprising frequencies in the audible spectrum. Using AM, the audio signals being transmitted by different stations are each shifted to different frequency bands, each centered about the frequency indicated on your radio dial. In this way, signals from different stations do not interfere with each other, which would cause distortion in the received signal. If modulation were not used, then all radio signals would be transmitted at the same time. At best, you would only be able to understand the station for which the signal received at your house is "loudest." At worst, the signals from all stations would simply add together, creating an incredible cacophony of dissonant sounds—which you would be unable to turn off.

AM is a subset of the group of modulation techniques that are said to be linear. Such techniques include double sideband modulation, upper sideband modulation, lower sideband modulation, and AM. All of these linear modulation techniques are similar, in that they shift the input signal up to a transmittable frequency by modulating the original signal using a sinusoid; differences in the bandwidth and frequency bands over which these methods are used result in important differences in implementation and application. It is important that you be aware that the process is more complex than the simplified explanation presented here, but a discussion of these methods is beyond the scope of this assignment. You should take a course in communication systems if you wish to further explore these topics.

In AM, the carrier signal is transmitted along with the information in the signal, allowing for demodulation (i.e., recovery of the audio signal) using low-cost envelope detection circuitry. The input signal is scaled to have a minimum value of no less than -1. The formula for AM is given by Equation 1 below

1. Amplitude Modulation Formula $$x_c(t) = A_c[1 + ax(t)]\cos(\omega_c t)$$

where A_c is the amplitude of the carrier signal, a is the modulation index—which scales the signal to make sure that its minimum value is approximately -1, and ω_c is the frequency of the carrier signal.

Consider the modulation of a simple waveform $2\cos(4\pi t)$ using a 100-Hz carrier wave. A modulation index of 0.5 provides the appropriate scaling. If we let the carrier amplitude be 1, then the modulated output signal is as shown in Figure 12.5.3.

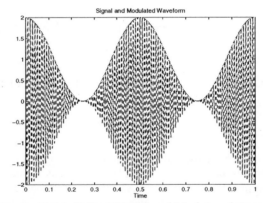

Figure 12.5.3 *Shifted Signal and Modulated Output*

You can see both the carrier and modulating waveforms clearly. Notice that the envelope of the modulated signal is the shape of the original signal, scaled by the modulation index with a constant shift of 1 on the vertical axis.

We can demodulate this signal by multiplying by a second sinusoid at the carrier frequency, $\cos(\omega_0 t)$, which results in a signal $r(t) = x_C(t)\cos(\omega_0 t)$. When expressed in terms of the original audio signal, it can be seen that $r(t)$ includes a $\cos^2(\omega_0 t)$ term. Substituting the trigonometric identity $\cos^2(\omega_0 t) = [1 + \cos(2\omega_0 t)]/2$ into this expression for $r(t)$ results in

---

Laboratory Assignment 12

$$r(t) = \frac{1}{2}A_c[1 + ax(t)][1 + \cos(2\omega_0 t)]$$

From $r(t)$, we can recover the original audio signal $x(t)$ by filtering out the high-frequency sinusoid at $2\omega_0$, assuming that ω_0 is large) using a LPF having a gain of 2. While the theory suggests that this method of demodulation will work, in practice it is often difficult to exactly match the carrier frequency and phase as it may drift during transmission and such circuitry is expensive.

Recall that the envelope of the modulated signal corresponds to the original signal. In practice, a simple, low-cost envelope detection circuit, like that shown in Figure 12.5.4, is often used to recover the original signal waveform from the modulated signal.

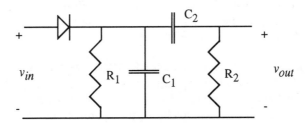

Figure 12.5.4 *Passive Envelope Detection Circuit*

The diode, R_1, and C_1 act as an envelope detector by allowing the peaks of the waveform to charge the capacitor, which does not completely discharge before the next peak occurs. It operates in a similar manner to a half-wave rectifier, except that the presence of R_1 allows the capacitor to discharge and thus follow a changing waveform. R_2 and C_2 form a high-pass filter to remove the DC component $1/2A_c$, which is a result of transmitting the carrier with the modulated signal.

The envelope detection circuit can operate as long as the carrier frequency is much higher than the highest frequency present in the signal. If the lowpass filter formed by R_1 and $C_1$is set to cut out signals higher than the highest signal frequency, then the carrier tone (which is significantly higher) should disappear. Similarly, the highpass filter formed by R_2 and C_2 should cut out the DC component of the signal.

Since the diode is a nonlinear device, it's difficult to simulate this circuit with standard linear operations. In MATLAB, the easiest way to perform this operation is to simulate the operation of the diode on the R/C circuit: if v_{in} is larger than the capacitor voltage, the capacitor voltage is charged to v_{in}. If v_{in} is smaller, the capacitor voltage discharges through R_1. This is easily accomplished using **for** loops and an **if** statement.

12.6 SOFTWARE EXPERIMENT

You are to do two things:

       1. Successfully simulate AM in MATLAB without having any aliasing occur;

       2. Demodulate an AM signal successfully in MATLAB, again without any aliasing.

These are nontrivial problems. You will need to know the frequency spectrum of the input signal, understand how modulation affects the signal, and be able to fix any problems that develop. Beware of aliasing. Remember that long signals require a great deal of computation, so be patient with your computer—it's thinking as fast as it can.

Problem 1.   Modulate the signal $x(t) = \cos(10\pi t)$ using a carrier frequency of 100 Hz. You are allowed to choose the sampling rate. The carrier amplitude should be 1. Listen to the output waveform and plot the frequency spectrum.

Problem 2.   Modulate the signal found in the file **P_12_1.wav** using an 8 kHz carrier; note that this carrier frequency is in the audible spectrum. The signal is a sample of a person talking. Try listening to the signal. Determine the signal frequency spectrum and eliminate any aliasing. You will have to change the sampling rate while leaving the signal integrity intact.

Problem 3.   Demodulate the signal you created in Problem 1. Make sure the output matches the input (except for amplitude). You may either emulate the behavior of the envelope detection circuit or the mathematical/filter technique.

Problem 4.   Demodulate the signal found in the file **P_12_2.wav**. You should determine the carrier frequency and bandwidth of the signal before applying any demodulation techniques to it. Listening to the demodulated output should help you to know when you have achieved the correct results, as the result should be comprehensible.

12.7 HARDWARE EXPERIMENT

You are to demodulate an AM signal successfully using the envelope detection circuit. You will need an amplifier stage following the demodulation so that you can listen to the audio output with a pair of headphones.

Problem 5.   Demodulate the signal you created in Problem 1. Play the modulated signal and feed the output into your envelope detection circuit. Look at the output with an oscilloscope—you should see a 5-Hz cosine waveform.

Problem 6.   Demodulate the signal found in the file **P_12_2.wav**. Play the modulated signal and feed the output into your envelope detection circuit. Listen to the output of your demodulation—it should be coherent.

Appendix A
Example Hardware Setup

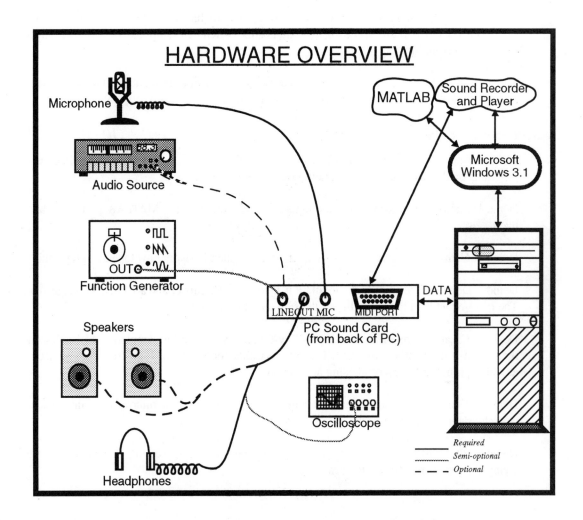

HARDWARE OVERVIEW

Microphone

Audio Source

OUT
Function Generator

Speakers

Headphones

LINE OUT MIC   MIDI PORT
PC Sound Card
(from back of PC)

Oscilloscope

MATLAB   Sound Recorder and Player

Microsoft Windows 3.1

DATA

Required
Semi-optional
Optional

This diagram shows an example laboratory setup using an IBM PC and associated software and hardware. The input sources are a microphone, a function generator, and an audio source like a CD player or tape deck. The output devices are speakers, headphones, or an oscilloscope. The only piece of additional PC hardware is a sound card with a line-in jack, a line-out or headphone jack, and a microphone jack. Optional sources are connected with a dashed line, and semi-optional sources (meaning that most lab experiments can be done without them) are connected with a semi-solid line.

MATLAB can interact with the sound card using Windows API functions, and audio signals can be sampled and played back using separate Windows programs. You may use either the standard Windows programs, or programs that are shipped with the sound card, or third-party software—it doesn't really matter, as long as they support recording and playback.

Our laboratory hardware at Carnegie Mellon University includes 486DX33 based PCs with Media Vision's Pro-Audio Spectrum/16 sound cards, which can sample at 44.1 kHz in stereo with 16-bit resolution.

The only special connector needed is one that can convert from the sound card mini-jack (the usual standard) to the rest of the world. We use a mini-jack to alligator clip cable.

Demonstrating the Hardware

As an example of using the sound card and associated hardware, we have included six demo programs on the distribution disk. They should be self-explanatory.

Demo 1: Sample a sine wave from the function generator, load into MATLAB, plot, and play back.

Demo 2: Using the sine wave from Demo 1, create a linearly rising envelope and play back.

Demo 3: Demonstrate amplitude modulation, using a 1-kHz carrier and a 100-Hz message tone. The real impact is in the audibility of the sum and difference tones.

Demo 4: Create discrete-time aliasing. A low-frequency sinusoid is sampled at 44.1 kHz and then down-shifted without incident. The process is then repeated for a high-frequency sinusoid, at which point the effects of aliasing are obvious.

Demo 5: Generate an echo. A person's laughter is recorded with the microphone, and an echo is synthesized and played back.

Demo 6: Pitch shifting by changing the sample playback rate: time scaling. A person's voice is changed from normal to fast to slow.

Appendix B
M-Files for Sound File Processing

There are many ways to store audio data. The most common ways, however, are "sound files," binary data files that contain some header information and the sampled data stored in a particular format. Some formats include signed 8-bit data, unsigned 16-bit data, and µ-law encoded data. MATLAB has some capability to parse these sound files, but it cannot handle every different type.

We have modified a publicly available program, SOX, which can convert over 15 different types of sound files from one to another, to include MATLAB .mat files. In addition, we have written several M-files that interface with SOX and load data from sound files.

SOX has been compiled for IBM PC platforms and is included on the source disk as sox.exe. Source code is also available for any other system, and should compile without difficulty.

The M-files are:

- **readwav**: Read Windows 3.1 .wav files, either 8 or 16 bits.

- **writewav**: Write Windows 3.1 .wav files, either 8 or 16 bits and any sampling rate.

- **readau**: Read Sun/DEC .au files, which are µ-law encoded.

- **writeau**: Write Sun/DEC .au files, which are µ-law encoded.

- **readsnd**: Read NeXT .snd files, which are µ-law encoded.

- **writesnd**: Write NeXT .snd files, which are µ-law encoded.

Each **read** M-file operates as a function and returns a column vector containing the samples from the sound file. The name of the file is given as a string argument; it must be in single quotes and have no extension (e.g., **readwav('P_2_1')** to read the file P_2_1.wav).

Each **write** m-file takes four arguments: the filename, the vector to be stored, the sampling rate, and the data size of the data contained in the vector—either 8 or 16 bits for .wav files, 8 bits only for .au or .snd files due to the encoding. E.g., **writewav('test1',x,8000,8)** to write out the vector **x** using an 8 kHz sampling rate and 8 bits of data.

Warning: SOX does no normalization of the data, except for bit shifting if necessary. Since MATLAB files can have several data types, including floating point, you must normalize your data before writing out to a file. All floating-point data types are rounded to integer values before storage, so it's important to make sure all values fit in the allowable data range (−127 to +127 for 8-bit data, −32767 to +32767 for 16-bit data).

SOX itself can support many different file types; if you need to add an M-file to support your system's sound file, the **readX** and **writeX** files are easily modified.

An excerpt from the SOX manual page:

FILE TYPES
  Sox needs to know the formats of the input and output files. File formats that have headers are checked; if that header doesn't seem right, the program exits with an appropriate message. Currently, the raw (no header), IRCAM Sound Files, Sound Blaster, SPARC .AU (w/header), Mac HCOM, PC/DOS .SOU, Sndtool, and Sounder, NeXT .SND, Windows 3.1 RIFF/WAV, Turtle Beach .SMP, and Apple/SGI AIFF and 8SVX formats are supported.

  .aiff AIFF files used on Apple IIc/IIgs and SGI. Note: the AIFF format supports only one SSND chunk. It does not support multiple sound chunks or the 8SVX musical instrument description format. AIFF files are multimedia archives and can have multiple audio and picture chunks. You may need a separate archiver to work with them.

  .au SUN Microsystems AU files. There are apparently many types of .au files; DEC has invented its own with a different magic number and word order. The .au handler can read these files but will not write them. Some .au files have valid AU headers and some do not. The latter are probably original SUN μ-law 8000-Hz samples. These can be dealt with using the .ul format (see below).

  .hcom Macintosh HCOM files. These are (apparently) Mac FSSD files with some variant of Huffman compression. The Macintosh has wacky file formats and this format handler apparently doesn't handle all the ones it should. Mac users will need your usual arsenal of file converters to deal with an HCOM file under UNIX or DOS.

  .raw Raw files (no header). The sample rate, size (byte, word, etc.), and style (signed, unsigned, etc.) of the sample file must be given. The number of channels defaults to 1.

  .ub, .sb, .uw, .sw, .ul These are several suffixes that serve as a shorthand for raw files with a given size and style. Thus, ub, sb, uw, sw, and ul correspond to "unsigned byte," "signed byte," "unsigned word," "signed word," and "ulaw" (byte). The sample rate defaults to 8000 Hz if not explicitly set, and the number of channels (as always) defaults to 1. There are lots of Sparc samples floating around in μ-law format with no header and fixed at a sample rate of 8000 Hz. (Certain sound management software cheerfully ignores the headers.) Similarly, most Mac sound files are in unsigned byte format with a sample rate of 11025 or 22050 Hz.

  .sf IRCAM Sound Files. SoundFiles are used by academic music software such as the CSound package and the MixView sound sample editor.

  .voc Sound Blaster VOC files. VOC files are multipart and contain silence parts, looping, and different sample rates for different chunks. On input, the silence parts are filled out, loops are rejected, and sample data with a new sample rate are rejected. Silence with a different sample rate is generated appropriately. On output, silence is not detected, nor are impossible sample rates.

.auto This is a "meta-type": specifying this type for an input file triggers some code that tries to guess the real type by looking for magic words in the header. If the type can't be guessed, the program exits with an error message. The input must be a plain file, not a pipe. This type can't be used for output files.

.smp Turtle Beach SampleVision files. SMP files are for use with the PC-DOS package SampleVision by Turtle Beach Softworks. This package is for communication to several MIDI samplers. All sample rates are supported by the package, although not all are supported by the samplers themselves. Currently, loop points are ignored.

.wav Windows 3.1 .WAV RIFF files. These appear to be very similar to IFF files, but not the same. They are the native sound file format of Windows 3.1. Obviously, Windows 3.1 is of such incredible importance to the computer industry that it just had to have its own sound file format.

.mat MATLAB 4.X .mat data files. MATLAB matrix data is stored in a binary format. Several different formats exist, from 16-bit words to double-precision floating point. The type of data is stored in the header. One note: if 8-bit data is saved to a .mat file, it's probably stored as a 16-bit word. Be careful—if you know your data is 8-bit, give a -b switch; if you know it's 16-bit, give a -w. The program will sort between the two types.

The echoplex effect is: Copyright (C) 1989 by Jef Poskanzer.
Permission to use, copy, modify, and distribute this software and its documentation for any purpose and without fee is hereby granted, provided that the above copyright notice appear in all copies and that both this copyright notice and this permission notice appear in supporting documentation. This software is provided "as is" without express or implied warranty.

Appendix C
Laboratory Report Guidelines

Every instructor has a different definition of the perfect lab report. We use the set of guidelines given below in our signals and systems courses; these are not meant as the definitive guide to writing lab reports, but should merely serve as a template if you do not already have guidelines in place. We suspect that instructors may wish to provide their own guidelines, particularly regarding laboratory performance evaluation.

Motivation

While many engineering students are adept at mathematics, problem-solving, and programming, fewer are as skilled at writing reports. The ability to accurately, concisely, and clearly maintain a record of your experimental activities and report on your results will be critical to your success as an engineering professional. You are more likely to be promoted if your employer is aware of the importance and extent of your activities; your work is more likely to be implemented in products if you can convince others of the value of your work and communicate its relevance; you are more likely to ultimately be promoted to a leadership position where you help define company technology directions, and thus work on projects which interest and excite you, if you are organized so that others can pick up on projects where you left off.

You are asked to provide reports for each laboratory assignment using the guidelines provided below. The objective is for you to become comfortable at both working in the laboratory environment and reporting your results. Each report should not be viewed as a major project, like a term paper, but as a progress report. Our goal is for your ability to evolve over the course of the semester so that you become more adept and efficient at maintaining an accurate record and reporting your results.

Report Goals

Following are some objectives of lab reports, which you should recall when writing and organizing your results as well as when proofing and evaluating your report. In general, a lab report has four dual purposes:

1. *To maintain a complete record of what you understood and accomplished.* Use figures, sketches, and plots, in addition to verbal and mathematical descriptions, to illustrate key ideas critical to your understanding. Documenting your work from as many different representations as possible aids in enhancing your understanding and clarifying your accomplishments for a wider audience. Identify design parameters selected and design procedures and equipment settings used. Indicate any difficulties encountered and your solutions to them. Given your report, you should be able to—at some later date—use your own report as a reference to help you do related work. In evaluating your report, can you do this?

2. *To provide for colleagues or new employees on the project a guide to what has been done, how it was accomplished, and the tools available.* Consider whether someone knowledgeable of the overall goals of a project could pick up your report, understand how and what you accomplished, and be able to determine the tools and information needed to replicate your results. Think of this objective in terms of a fellow student who takes this course in a later semester: if assigned a similar experiment, could this student use your report to help her/him gain some understanding and get started? You do not want to rewrite the lab assignment, since it can be referenced for details. However, your report should be self-contained, and understanding it should not require simultaneous access to the assignment.

3. *To highlight key results and bottlenecks for your manager* including a critical evaluation of the results, recommendations for future work, and examples (either simple analytical or experimental results with figures) illustrating key points. Using your report, could a manager make informed decisions regarding what is the most reasonable direction for the project to take at this point? Would your manager think that you had done a thorough and thoughtful job? Would your manager believe that you had creative potential and good ideas?

4. *To demonstrate your understanding for your manager,* or—for this course—to demonstrate that you completed and understood the laboratory assignment. Essentially, if you are able to write a report using your own words such that it meets the first three goals, then you will have demonstrated the required understanding.

General Format Guidelines

In completing your final report, make certain to adhere to the following guidelines:

• It should not be lengthy, but should contain enough information to meet the goals stated above. We have found that 5-10 pages of text, excluding figures and other attachments, is generally sufficient for the purposes of these assignments.

• Figures, plots, and tables should be clearly identified. Be sure to label all axes and highlight points of special interest, including critical component parts and interconnections. We encourage you to use visual explanations when possible, but caution you that such visual aids are not self-explanatory and still need to be discussed in the text. Also, do not simply attach plots of all the results that you generated. Be selective: include only those figures that best illustrate your main points.

• As part of your text, include answers to the questions posed in the assigned laboratory problems. Be sure that the correspondence between the assigned tasks and questions and your answers is clearly evident from the text.

• You may include printouts of the MATLAB function files you created for the assignment if they are referred to in the text of your report, but it is not required. In fact, we discourage you from attaching MATLAB files unless including them enhances your discussion.

• Attach your final, corrected solutions to assigned lab prep problems. In explaining your results, you may wish to reference these solutions rather than rewriting explanations.

Suggested Report Organization

Below is a suggested outline for your report. You are welcome to use a different organization if you feel that it would result in a clearer presentation of the material. If you do use a different organization, be sure that you include all the information requested in the outline below.

1. **Title Page** should include

   • Laboratory title and assignment number
   • Group member names
   • Course and section numbers
   • Date

2. **Introduction**

   • Approximately 1/2 page
   • Express the goals and assigned tasks for the laboratory in the application context using your own words. What are you given? What are you to find?
   • What are the applicable analytical representations, concepts, operations, and tools you have learned from the text and lectures?
   • What did you learn from this assignment?

3. **Mathematical Analysis**

   • Approximately 1 page of text, with references to attached Laboratory Prep Problem solutions
   • Define each given and desired quantity as a variable.
   • Present analytical (mathematical) formulations of the problems you are to solve and tasks you are to perform.
   • Provide mathematical, intuitive, and physical explanations for what you expect to see in the assignment. You may want to use a simplified example that illustrates your expectations. For example, explore the analytical impact on a simple sinusoid of the operations that you will perform on speech in lab.
   • Present any preliminary analytical preparation needed to execute the laboratory assignment and identify key design parameters or variables that you will have to choose
   • Often the two tasks above are facilitated by referencing your Lab Prep Problem solutions. How do these solutions relate to the laboratory goals, tasks, and expected results?

4. **Experimental Parameters and Method**

   • Approximately 1 page of text, with references to diagrams as necessary

• Identify the hardware systems or components that you will be using. Be specific; e.g., what specific functionality of the sound card will you be using? Indicate how the hardware will be interconnected to perform the assigned tasks, and identify hardware settings that must be selected as part of the data measurement or acquisition process. Which components, settings and operations will be most critical? Most sensitive?

• Which MATLAB functions and commands will you be using? What function do they perform? Identify how the variables that you defined in the Analysis section correspond to parameters and signals that will be passed to and returned from these MATLAB functions.

• Often a diagram is helpful for specifying the interrelationships among the hardware, software, variables, parameters, and settings.

5. Results

• Approximately 1-2 pages of text, with references to plots as necessary

• How did you complete the assigned tasks? Indicate signals and systems representations, concepts and operations used, equipment settings selected, and parameters computed. What procedure did you use for selecting successful settings for the equipment and parameters for MATLAB functions?

• What did you observe? Document your observations in as much detail as possible. You should comment on the interplay between the mathematical analysis and observed physical properties. Use the questions in the laboratory as guidelines in your discussion, but do not limit yourself to asking only these questions.

6. Evaluation and Conclusions

• How good were your experimental results? In what ways did they match with your expectations? Explain your answers. Suggest ways to improve your results.

• What difficulties did you encounter, if any? Be sure to document how you overcame these. Even difficulties that seem to have trivial solutions, e.g., MATLAB syntax errors, should be documented so that you do not continue to repeat them in future labs.

• Which tasks were easiest to analyze, understand, and execute? Most difficult? Why? If some major difficulty occurred that did not allow you to complete the assigned task in the allotted time, analyze what went wrong and indicate how you expect to avoid similar difficulties in future labs.

• What extensions come to mind? What else might you like to try, given the time and resources?

Grading

Your laboratory reports will be graded based on

- Content: You should have completed all of the assigned tasks and should demonstrate a clear mathematical and physical understanding of your observations.

- Presentation: Your report should be as clear and concise as possible. All discussions should be in your own words and reflect your observations, experiences, and understanding

- Use of Visual Aids: Since you are an engineering student, you should use figures, plots, diagrams, and tables to represent information in a clear and concise form. However, do not simply attach all figures generated in the course of your experiment; be selective. All attached visual aids must be referenced in the text discussion and be included only if they enhance the clarity of the report.

- Observations: You need to demonstrate that you were careful in your observations. Provide quantitative detail. Previous observations should be used to help define questions to be answered and which experimental parameters to change.

- Conclusions: You should demonstrate that you thought carefully about how your mathematical analysis, experimental observations, and physical intuition match for the given problem. What discrepancies do you notice? How can you explain these?

Laboratory Evaluation Form Template

Laboratory Assignment No. _____   Title _____

Lab Group Names _____ Lab Prep % _____

_____ _____

_____ _____

Score:

_____ Final lab prep solutions
(20)

_____ Laboratory Demo
(40)
    • Instructor-provided list of assigned problems

    • Instructor/Assistant initialize upon successful demonstration and
      Note any unusual (good or bad) performance.

_____ Laboratory Report
(40) • Graded based on criteria described above

_____
Total

Index